Principles of Plant Culture

Principles of Plant Culture

E. S. Goff

2023
Agri Horti Press
New Delhi-110 084

ISBN: 9789392969225 (Hardbound)

Published By

Agri Horti Press

New Delhi

E-mail: agrihortipress@gmail.com

Digitally Printed at : Replika Press Pvt. Ltd.

PREFACE

This book has grown out of the author's experience in the lecture room and laboratory, while giving instruction to students in the Short Course in Agriculture, in the University of Wisconsin.

It is intended especially for students who have had little or no previous instruction in botany, and it is hoped that it may also be found interesting and profitable to the general reader who would learn more of the principles that underlie the culture of plants.

It is expected that the instructor will amplify the text in proportion to the time at his command, and the capacity of his students. In the author's practice, the first three chapters have been found sufficient for a term of twelve weeks, and the remaining chapters, supplemented with some special work in horticulture, have served for a second term.

A syllabus of laboratory work is added as an appendix.

It is hoped that this book may prove as useful to other instructors as it has proved to the author during its evolution.

E. S. GOFF.

NOTE.—In the second edition (1899) a number of verbal changes have been made and a few additional paragraphs and illustrations have been inserted.

ACKNOWLEDGEMENTS

The author desires to express his thanks to Prof. Charles R. Barnes, of the University of Chicago, Prof. F. W. Woll, of the Wisconsin Agricultural Experiment Station and Mr. F. Cranefield, the author's assistant in horticulture, for revision of the manuscript (of the first edition), and for many valuable suggestions.

Figures 14, 15, 18, 19, 20, 21, 22, 28, 29, 30, 34, 46, 47, 48, 49, 53, 54, 55 and 83 were copied from "Agricultural Botany," with the sanction of the author, Prof. M. C. Potter, of the Durham College of Science, England. Figures 61, 62, 63, 82, 84, 94, 98, 100, 101, 102, 103, 113, 117, 118, 119, 120, 121, 123 and 128 are from "The Nursery Book," by Prof. L. H. Bailey, and are used by permission. Figures 31, 59, 81, 85, 99 and 130 are from "The Amateur Fruit Book," by Prof. S. B. Green, and are used by permission. Figure 32 is from a photograph kindly loaned by Prof. Green. Figures 37, 39, 40, 41, 42 and 44 are copied by permission of the publishers from "Barry's Fruit Garden." Figures 94 and 95 are from "How to Make the Garden Pay," by T. Greiner, and are used by permission. Figures 65 and 66 were copied by permission from Bulletin No. 37, of the Rhode Island Agricultural Experiment Station. Figures 77, 78 and 79 are from cuts in the possession of the Wisconsin Agricultural Experiment Station.

CONTENTS

PRINCIPLES OF PLANT CULTURE

CHAPTER I

INTRODUCTORY

Before taking up a systematic study of plant culture, we may profitably consider a few principles of a more general nature.

1. Close Observation offers the best means of gaining knowledge of material things. The habit of accurate discernment, and of studying the relations of and the reasons for things and facts as we find them, should be constantly cultivated. Knowledge once gained must be applied in the proper *place*, the proper *manner* and at the proper *time*, or the highest success in any calling cannot be expected.

2. The Difference between **Art** and **Science.** Art is simply knowing how to do a thing without reference to reasons. Science considers the reasons for doing it in a particular manner. Art implies more or less of skill gained through practice. Science implies a knowledge of the objects to be gained by a given operation and the conditions affecting the process.

An intelligent but ignorant person might be taught to prepare and insert a cion (386)* in the most approved

* The numbers in parenthesis in the text refer to the numbered paragraphs in this book, and are intended to help students to a better understanding of the subject. Students should be urged to look up these cross references.

manner. This pertains to the *art* of grafting. The same person might be taught the *reasons why* each step of the process is performed in its particular manner. This pertains to the *science* of grafting. One may become a skilled grafter without learning the science of grafting, but he cannot graft intelligently. The artisan, however skillful, who knows only the art, cannot become a master workman in the highest sense until he learns also the science that underlies his trade.

The art of doing any kind of work is best learned by working under the guidance of a skilled workman. The science is best learned from books with the help of trained instructors. Science not yet wrought out, and hence not explained in publications, is learned by close, persistent and thoughtful observation and study.

3. Environment is a term used to express all the outside influences, taken as a whole, that affect a given object in any way. A plant or animal, for example, is affected by various external conditions, as heat, moisture, light, food etc. These conditions and all others that influence the plant or animal make up its environment.

4. What is Culture? The well-being of a plant or animal depends very much upon a favorable condition of environment, and with the proper knowledge, we can do much toward keeping the environment in a favorable condition. For example, if the soil in which a plant is rooted lacks plant food, we can enrich it; if it lacks sufficient moisture, we can dampen it; if the plant is shaded by weeds, we can remove them. These, and any other things that we can do to make the environment more favorable, constitute *culture* in the broadest sense of the term. A full knowledge of the culture of any plant implies a knowledge, not only of the plant and its needs, but of each separate factor in its environment, and how to maintain this factor in the condition that best favors the plant's development toward some special end, as the production of the finest and highest type of fruit, flowers or seed. We should know, not only the soil that best suits the plant, but the amount of light, moisture, warmth and food in which it prospers best. We should know the enemies that prey upon it, the manner in which they work their harm, and how to prevent their ravages. We should know, in short, how to regulate every factor of environment so as to promote the plant's well-being to the utmost, as well as how to develop every desirable quality the plant possesses.

5. Domestic or **Domesticated Plants** or **Animals** are those that are in the state of culture. In nature, different plants and animals struggle with one another for space and food. Only those best adapted to their environment survive, and these are often much restricted in their development. In culture, the intelligence and the energy of man produce a more favorable environment for the species he desires to rear; hence domestic plants and animals attain higher development in certain directions than their wild parents. The cultivated potato, for example, grows larger, is more productive and is higher in food value than the wild potato. The finer breeds of horses and cattle are superior to their wild progenitors in usefulness to man.

6. Culture Aims to **Improve Nature's Methods** rather than to imitate them. By cutting out the superfluous branches from a fruit tree, we enable the fruit on the remaining branches to reach a higher state of development. By planting corn at the proper distances, we prevent crowding and enable each plant to attain its maximum growth. We should constantly study nature's methods for useful hints in culture, and the culture of a given plant or animal should be based more or less upon its natural growth conditions, but the highest progress would be impossible if we sought only to imitate nature.

7. Culture Deals with **Life.** All the products of culture, whether obtained from the farm, garden, orchard, nursery or greenhouse, proceed directly or indirectly from plants or animals, both of which are living beings. A knowledge of the conditions that sustain and promote life, is, therefore, the foundation to a broad knowledge of husbandry.

8. What is Life? We know nothing of life except as it is manifested through the bodies of plants and animals. With these, we can define, within certain limits, the range of environment in which it can exist; we can hinder or favor it; we can apparently destroy, but we cannot restore it. We know that it proceeds from a parental body similar to its own, that the body it inhabits undergoes a definite, progressive period of development, at the end of which the life disappears and the body loses more or less promptly its form and properties.

9. Vigor and **Feebleness** are terms used to express the relative energy manifested by the life of different living beings. Certain trees in the nursery row usually outstrip others in growth, i. e., are more vigorous than others.

One pig in a litter very often grows slower than any of the others, i. e., is more feeble or less vigorous than any of the others. Feebleness is the opposite of vigor. The most vigorous plant or animal usually attains the largest size, and as a rule, is most satisfactory to its owner. Vigor is promoted by a favorable environment. It is usually greatest in rather young plants and animals, and declines with advancing age. It may be reduced by disease or improper treatment, and when thus reduced is often difficult to restore. Reduced vigor tends to early maturity and shortened life, and sometimes to increased prolificacy.

10. Hardiness and **Tenderness** are terms used to express the relative power possessed by different plants or animals to endure extremes in their environment. The Oldenburgh apple endures with little harm vicissitudes of temperature that are fatal to many other varieties; in other words, it is hardier as regards temperature than many other varieties. The reindeer is hardier as regards cold than the horse, but tenderer as regards heat. The melon plant is hardier as regards heat and drought than the lettuce, but tenderer as regards wet or cold.

11. Health and **Disease.** A plant or animal is said to be *in health* when all its organs (parts) are capable of performing their normal functions. An organ incapable of doing this, or the being possessing such an organ, is said to be *diseased.*

12. The Cellular Structure of **Living Beings.** A bit of vegetable or animal substance, examined under a microscope of moderately high power, is seen to be made up

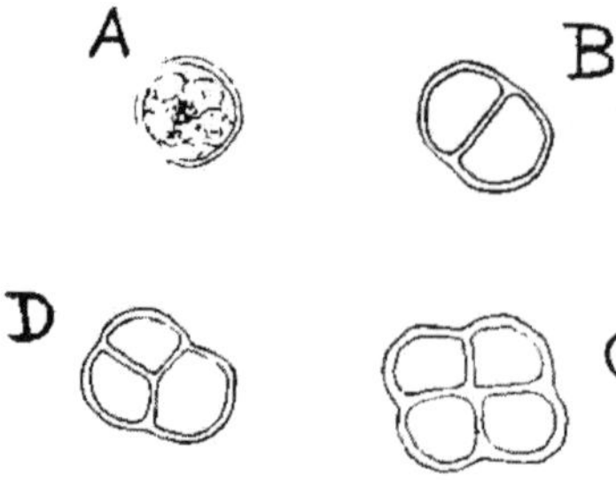

FIG. 1. Showing four individual plants of a species of *Protococcus*. A shows a plant before commencing to divide into other plants. B, C and D show how the cells divide to form other plants. Highly magnified.

of numerous little sacks or cavities, more or less clearly defined, called *cells*. Cells from different beings, or from different parts of the same being, may vary much in form and size, but they are seldom large enough to be seen without magnifying power. Some of the lowest plants and animals consist of single cells (Fig. 1). Some of the lower plants consist of a single row of cells united at the ends (Fig. 2). The higher plants

FIG. 2. Part of a filament of a species of *Spirogyra*, a plant consisting of a single row of cells united at their ends. The places where the cells join are indicated by the vertical lines. Highly magnified.

and animals are made up of many cells united, and in these, the cells assume different forms and properties in the different organs (Fig. 3). In some cases the united cells may be readily separated from one another, which shows each cell to be more or less an independent structure. As a rule, each cell is surrounded by its own closed cell-wall.

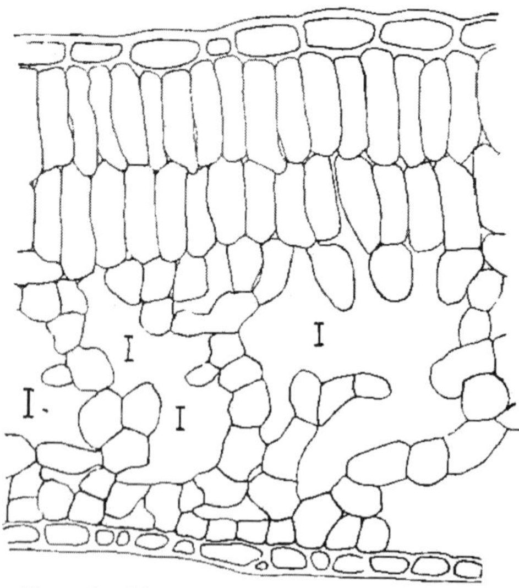

FIG. 3. Showing cells of the apple leaf in a section from its upper to its lower surface. Highly magnified. The spaces marked I are cavities between the cells.

13. Protoplasm (pro′-to-plasm). Living cells consist of a transparent, jelly-like substance called *protoplasm*, which manifests the various phenomena of life. Protoplasm may exist either in an *active* or *dormant* state. In the active state it requires both nourishment and oxygen; in the dormant state it may exist for a considerable time with very little of either, and is far less susceptible to external influences than in its active state. The protoplasm contained in plants during their rest period (171), in mature air-dry* seeds, and in the lower animals during their torpid condition, is in the dormant state.

14. Reserve Food. Active protoplasm may absorb nourishment in excess of immediate requirements and hold it as reserve food. In plants, this reserve food is in the form of starch, sugar or oil; in animals it is in the form of fat. These substances are formed by the protoplasm from its crude food materials (59). The reserve food enables the plant or animal to live through limited periods of scarcity, and to meet the demands necessitated by reproduction (16).

15. Growth is the normal, permanent change in the form of a living vegetable or animal body, and is usually accompanied by increase in size. It may occur either through expansion of cells already formed, or through cell multiplication. The latter may take place either by division of older cells into two or more smaller cells (Fig. 1), or by the formation of new cells within older ones — the young cells thus formed attaining full size by subsequent enlargement.

---

* Material is said to be "air-dry" when it is as dry as it will become by exposure to the air at ordinary temperatures.

16. Reproduction is the increase in number of living beings. It is one of the properties of protoplasm and is essentially a process of division. As living cells multiply by forming new cells, so living beings, which consist of cells, multiply by the separation of a part of their own cells, and this separated group of cells grows into a complete organism like the parent. The higher plants multiply by seeds (156), which are separated from the parent plant, and each of which contains a young plant (54). The eggs from which young birds are hatched contain cells filled with living protoplasm, and the protoplasm of the living young of mammals is separated from the parent before birth.

17. Reproduction is either **Sexual** or **Non-Sexual.** *Sexual* reproduction can take place, as a rule, only upon the union of cells of different sexes. It is not peculiar to the animal kingdom, but occurs in plants also, and except in rare cases, is necessary to the production of seeds that are capable of germination (28). It is the only method of reproduction in the higher animals. Sexual reproduction does not usually take place until the period of most rapid growth has passed. *Non-sexual* reproduction is independent of sex. It results from the direct separation of a part of the parent, which under favorable conditions develops into a complete individual. It occurs when plants multiply by means other than by seeds, as by non-sexual spores (53), bulbs (352), stolons (348), cuttings (358), etc., and it is a common method of reproduction in certain of the lower animals, as plant lice (aphidæ).

18. Heredity and **Variation.** The offspring of a plant or animal tends to be like the parent or parents. But no two beings can be begotten and developed in exactly the same environment, and since environment always affects the individual more or less, it follows that no two individuals can be precisely alike. Variation in the offspring may take place in any direction, as in the size or color of the flower, the sweetness or juiciness of the fruit, the prolificacy, the vigor (9), or the hardiness (10) etc. It follows that in culture certain individual plants or animals are more desirable to the cultivator than others, because the individuals possess different qualities.

19. The Principle of Selection. Since the offspring tends to resemble the parent or parents, we may gradually improve plants or animals in the direction of greater usefulness by selecting the most desirable individuals for reproduction. For example, by saving and planting seeds from the plants that produce the finest petunia or pansy blossoms, we secure finer flowers than if we gather seeds without regard to parentage.

20. Breeding in plants and animals is reproduction, watched over and directed by man, with reference to securing special qualities in the offspring. It is based on the principle that the peculiarities of the parent or parents tend to be reproduced, and may be intensified, in the descendants. But before we are prepared for the study of breeding, we need to know something of the principles of classification.

21. Classification is the arrangement of the different kinds of plants and animals into groups and families based on individual resemblances. If we examine plants as they are growing in nature, we may observe (*a*) that there are many plants of the same kind, and (*b*) that there are many kinds of plants. The different plants or animals of the same kind are called *individuals*, and, in general, we may say that the different kinds of plants or animals are called *species* (spé-cies). But these simple distinctions are not sufficient to satisfy the needs of natural history. We might say, for example, that the violet is one kind of plant and the oak is another, which is true; but there are also different kinds of violets and different kinds of oaks. We might say that the apple is one kind of plant and the pear is another, but there are different kinds of apples, as the crab apple and the common apple, and there are different kinds of crab apples, and of common apples. We must not only arrange the kinds of plants into groups, but we must have groups of different grades. For example, botanists call each distinct kind of plant, as the sugar maple, the white oak and the dandelion, a *species*. Then the species that rather closely resemble each other are formed into groups, each of which is called a *genus* (ge′-nus), plural, *genera* (gen′-e-ra) as the sugar maple and the soft maple; the white oak, the red oak and the bur oak; the raspberry and the blackberry; and the apple, pear and quince. Then the genera that resemble each other, as the one containing the apple, pear and quince, and the one containing the plum, cherry and peach, are formed into other groups called *families.** Thus *families are made up of genera*, and *genera are made up of species.* There may be, also, different *varieties* in the same species, as the different varieties of apple, pea, or strawberry.

* Related families are often further united into orders.

An extensive retail bookstore furnishes an object lesson in classification, though we must remember that in natural history it is usually the names and descriptions of plants and animals that are classified, and not the plants and animals themselves. In the bookstore, we will observe that the books are not placed upon the shelves without order, but that they are arranged in groups. Different copies of the same work are placed together. Different works on the same subject, as Gray's botany, Wood's botany, Bessey's botany are also placed in a larger group. Then all the scientific books are formed into a still larger group, as are the books of fiction, the books of poetry, the music books, etc. Comparing this arrangement with that employed in natural history, each separate work, as Gray's Manual of Botany, Thomas' Fruit Culturist, Bunyan's Pilgrim's Progress, etc., would correspond to a species,* and the different copies of the same work would correspond to individuals. The books treating of the same general subject, as the different works on geology, botany or arithmetic would correspond to genera, and the different classes of books, as scientific books, books of fiction, etc., would correspond to families. There would also be copies of the same work in different bindings which would correspond to varieties.

22. Scientific Names are given to plants and animals because the common names by which they are known are so often local. For example, quack grass, one of our common troublesome weeds, is known by at least seven

* It should not, however, be understood that the different species of plants and animals are always as readily distinguished as are the different works in a bookstore.

different common names in this country alone, and yet, in a given locality it is often known by only one name. Its scientific name, however, *Agropyrum repens* (a-gro-py'-rum re'-pens), is the same in all languages and countries. Scientific names are usually Latin and consist of two words. The first word is the name of the genus to which the plant or animal belongs, and is called the *generic* (ge-ner'-ic) name; the second word designates the species, and is called the *specific* (spe-cif'-ic) name. For example, *Pyrus malus* (py'-rus ma'lus) is the scientific name of common apple, *Pyrus* being the *genus* to which the apple belongs, and *malus* designating which *species* of the genus is meant.

23. Crosses and **Hybrids** (hy'-brids). We have seen that in sexual reproduction, a union of male and female cells is almost always essential (17). When these cells proceed from two individuals of different varieties (21, 436), the offspring is called a *cross;* when they proceed from individuals of different species, it is called a *hybrid.* Hybrids are possible only between closely-allied species and are often incapable of reproduction, in which case they are said to be *sterile.* The mule, which is a hybrid between the horse and the ass, is a familar example of a sterile hybrid. Sterile hybrids are not uncommon in plants. A hybrid that is capable of reproduction is called a *fertile* hybrid.

Hybrids and crosses may resemble both parents about equally or they may resemble one parent more than the other. They sometimes differ materially from either parent. The offspring of crosses and fertile hybrids is generally variable in proportion as their parents were different from each other, and this variability may continue through several generations.

24. The Theory of Evolution, now generally accepted by naturalists, assumes that the higher plants and animals have been gradually evolved from lower forms, through the principle that those individuals possessing peculiarities best fitting them to endure the adverse conditions of environment have survived and perpetuated their kind, while others have perished.

25. Parasites. Both plants and animals are subject to being preyed upon by other, usually smaller, plants and animals, that live upon or within their bodies, consuming the tissues of their bodies or their reserve food. Plants or animals that derive their nourishment from other plants or animals are called *parasites* (par′-a-sites) or *parasitic*. The plant or animal from which a parasite derives its nourishment is called a *host*. Parasites are often microscopic in size. They are generally more or less injurious to their host, and form one of the most fruitful sources of disease (271). Some, however, as the micro-organisms of the roots of clover and other leguminous plants, are beneficial (113).

CHAPTER II

THE ROUND OF PLANT LIFE

The earliest stages of plant growth occur in the seed, hence this is an appropriate place to commence our study. We will first consider

SECTION I. THE BEHAVIOR OF SEEDS TOWARD WATER

26. Seeds Absorb Water when placed in contact with it. If we fill a bottle with air-dry beans, then pour in all the tepid water the bottle will contain, taking care to shake out the air bubbles, and place the bottle in a warm room, the beans will soon swell until they have pressed each other quite out of shape, and no water will be forced out of the bottle. This shows that the beans have absorbed the water and have swollen in consequence. This quality of absorbing water by contact, at ordinary temperatures, is possessed to a greater or less extent by most seeds, and indeed by nearly all air-dry vegetable material. It is unnecessary that the seeds be covered with water to enable them to absorb it. If in contact with any moist medium, as a damp cloth or damp earth, they will absorb moisture and swell.

27. The Rate at which **Seeds Absorb Water** depends upon several conditions, as

a—*The water content of the medium* with which they are in contact. If we place one lot of beans in water, a second in wet earth and a third in slightly damp earth, we shall find that the first lot will swell fastest, the second next and the third slowest. Few seeds will absorb enough water from damp air at ordinary temperatures to swell much.

b—*The points of contact.* If we weigh, two lots of 100 beans each, on a delicate balance, and mix each lot with well-crumbled, moist loam in a fruit jar, packing the loam tightly in one of the jars and leaving it as loose as possible in the other, close both jars to prevent evaporation, and after twenty-four hours sift the beans out of the loam and weigh the two lots again—we shall find that the beans in the jar containing the compacted loam have increased more in weight than the others. This indicates that the beans in this jar have absorbed water faster than those in the other, because they were in contact with the moist loam at more points.

c—*Temperature.* If we fill two bottles with beans, adding ice water to one, placing it in a refrigerator, and lukewarm water to the other, setting it in a warm room, we shall find that the beans in the latter bottle will swell more rapidly than those in the former. This shows that a warm temperature favors the absorption of water—a fact that is true of all seeds. The same would have been true had we planted the beans in two samples of moist earth, placing these in different temperatures.

d—*The nature of the seed-case.** In the bean, Indian

---

* The term seed-case is here used to designate the outer covering of the seed as the word seed is understood by the seedsman or planter. Every seed, as we buy it in the market, or when ready for planting, has one or more covering layers. In the peanut, for example, what we here call the seed-case is commonly called the shuck; in the cocoanut it is called the shell; in the bean and Indian corn it is more often called the skin. In botany, the outer coverings of seeds are given different names, as peri-

corn, wheat and many other seeds, the seed-case is of such a nature that it absorbs and transmits water readily. In certain seeds, however, as of the honey locust, canna, thorn apple etc., especially if they have been allowed to become dry, the seed-case does not readily transmit water at growing temperatures. Such seeds may lie for weeks, and even months, in tepid water without swelling, but when the water is heated to a certain degree, they swell promptly, a fact often turned to account by nurserymen (36). We cannot always judge by the appearance of a seed-case whether it will transmit water readily or not.

Section II. Germination

28. What is Germination? If we place a few viable* grains of Indian corn between the moist cloths of a seed-tester (Fig. 5), cover with the glass and place in a warm room, we shall observe if we examine the corn frequently, that a change, aside from the swelling, will soon take place in at least a part of the grains. The seed-case will be burst by the pressure of a tiny white shoot from beneath. We say that such grains have sprouted or have commenced to *germinate* (ger′-mi-nate), i. e., have taken the first visible step toward developing into a plant.

We have seen that the mature seed contains protoplasm in its dormant condition (13). At a suitable temperature, the protoplasm, on the absorption of water, re-

---

carp, testa, etc., according to their exact office in the make-up of the plant. To avoid explaining the technicalities of a complex subject, it seems preferable to adopt a term that will include the various words used in botany to designate the outer coverings of seeds.

* A viable (vi′-a-ble) seed is one that is capable of germination. Not all seeds are viable (165).

sumes its active state, and the cells of a certain part of the seed begin to increase in size and to divide (15), causing the tiny shoot to burst through the seed-case. Germination is completed when the young plant (plantlet) is sufficiently developed to live without further aid from the seed.

29. Moisture is **Essential** to **Germination.** Air-dry corn or other seeds will not germinate if kept however long in a warm room, whereas viable seeds, that have absorbed water until fully swollen, will usually germinate if exposed to air of a suitable temperature, under conditions that prevent their loss of moisture. This shows that a certain amount of moisture must be absorbed by the seed before germination can take place. Seeds must be nearly or quite saturated with water before they will germinate.

In culture we *plant* seeds in some moist medium, usually the soil, in order that they may absorb moisture and germinate, and thus develop into new plants.

30. Warmth is **Essential** to **Germination.** Had we placed the seed-tester mentioned in paragraph 28 in a refrigerator in which the temperature never rises above 46° F., instead of in a warm room, the corn grains would not have germinated however long they remained there. This shows that a certain degree of warmth is also necessary to germination. Without this, the protoplasm ot the seed cannot assume its active state (13). The lowest (*minimum*) temperature at which seeds can germinate varies considerably with different species, and so does the temperature at which they germinate soonest (*optimum*) as also the highest (*maximum*) temperature at which they can germinate. The following table* shows approximately the minimum, optimum and maximum temperatures at which seeds of the species named germinate.

	MINIMUM.	OPTIMUM.	MAXIMUM.
Barley	41° F.	......77°–88° F.	...... 99°–111° F.
Bean (Scarlet runner)	41	......77 –88	...... 88 –99
Buckwheat	41	93	115
Clover (red)	88–99	......99 –111	......111 –122
Cucumber	60–65	......88 –99	......111 –122
Flax	41	......77 –88	...... 88 –99
Hemp	32–41	......99 –111	......111 –122
Indian corn	41–51	......99 –111	......111 –122
Lucern (Alfalfa)	88–99	......99 –111	......111 –122
Melon	60–65	......88 –99	......111 –122
Oat	32–41	......77 –88	...... 88 –99
Pea	32–41	......77 –88	...... 88 –99
Pumpkin	51–60	......93 –111	......111 –122
Rye	32–41	......77 –88	...... 88 –99
Sunflower	41–51	......88 –99	...... 99 –111
Wheat	32–41	......77 –88	...... 88 –108

These temperatures refer to the soil or other medium with which the seeds are in contact, and not to the atmosphere.

When moisture is sufficient, the time from planting to sprouting decreases rapidly as we approach the optimum temperature. In an experiment, Indian corn sprouted in one-third of the time at 88° F. that it required to sprout at 61°.

31. Free Oxygen is **Essential** to **Germination.** If we place in the bottom of each of two saucers† a layer of puddled‡

* Compiled from Haberlandt and Sachs.

† If flower-pot saucers are used they should first be well soaked in water, so that they will not extract water from the soil.

‡ Soil is said to be *puddled* when wet and packed until it is in the consistency of putty.

clay or loam, put 25 viable beans on the soil in each saucer, then fill one saucer with moist sand and the other with puddled clay or loam, pressing the latter down very closely around the beans, cover both saucers with a bell-jar, and place in a warm room for two or three days, we shall find that the beans covered with the sand will sprout promptly, while those covered with the puddled soil will not (Fig. 4). In the sand-covered saucer the

FIG. 4. In the left hand saucer beans were planted in puddled soil. In the other, they were covered with sand. They failed to germinate in the puddled soil, because their contact with oxygen was cut off. (From nature).

air between the grains of sand has had access to the beans, while in the other the air has been shut out, which explains the sprouting of one lot of seeds and the failure of the other. About one-fifth of the atmosphere is free oxygen, i. e., oxygen that is not chemically combined with any other substance.

We have seen that protoplasm in its active state requires oxygen (13). Unless seeds are so planted that a certain amount of this free oxygen can reach them they cannot germinate.* Ordinary water contains a little free oxygen, but not enough to enable many kinds of seeds to germinate in it, though the seeds of some water plants, as the water lily and rice will germinate in water. But even these will not germinate in water that has been

*This probably explains why very deeply-planted seeds rarely germinate.

boiled long enough to expel the oxygen, and is placed under conditions that prevent its absorption again (Fig. 5).

We thus see that seeds require three conditions before they can germinate, viz., a certain amount of moisture, of warmth and of oxygen. In planting seeds, we should consider all these requirements.

32. Prompt Germination is **Important.** As a rule, the sooner a seed germinates after it is planted, the better, for it is generally in danger of being destroyed by animals or fungi, and the plantlet probably loses vigor by too slow development. Weeds may also be gaining a start if germination is delayed. We should, therefore, treat both the seed and the soil in the way that favors prompt germination.

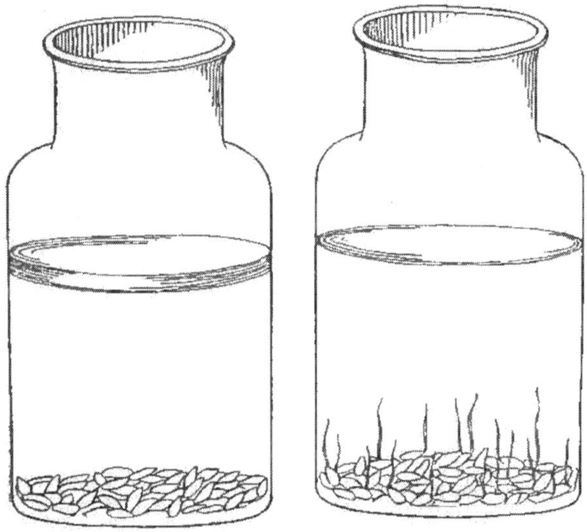

FIG. 5. In the left bottle, the water, which had been boiled to expel the oxygen, was covered with oil to prevent it from absorbing oxygen again, hence the rice seeds in it could not germinate. In the right bottle, the water was not covered, and so could absorb oxygen, permitting the seeds to germinate. From nature.

33. Compacting the **Soil** about planted seeds **Hastens Germination** by multiplying their points of contact with the moist earth (27*b*). When the soil is becoming drier day by day, as it often is in spring, compacting the soil about planted seeds materially hastens their germination and often secures germination that without the compacting might be indefinitely postponed. The hoe, the feet, a board or the hand- or horse roller may be used to compact soil over planted seeds.

34. Planting should be **Deferred** until the **Soil** becomes **Warm.** Seeds cannot germinate promptly until the temperature of the soil in which they are planted approaches the optimum for their germination (30) during the warmer part of the day, and germination is promoted little, if at all, by planting before this time.

35. Excess of **Water** in the soil **Retards Germination** by restricting the supply of oxygen (31), and sometimes, by keeping the soil cold. Seeds should not be planted in soil wet enough to puddle (31) about them, nor should the soil in which the seeds of land plants are planted be so freely watered that the seeds remain surrounded with liquid water, thus shutting out the normal supply of oxygen.

36. Germination may be **Hastened** by **Soaking** seeds before planting. Since seeds cannot germinate until nearly or quite saturated with water (29), and since they absorb water faster from a very wet than from a damp medium (27*a*), and in a warm than in a cool temperature (27*c*), we may hasten germination a little if the soil to receive the seeds is only slightly moist, by soaking the seeds before planting in warm or slightly hot water until they have swollen. This method is sometimes practiced by gardeners with sweet corn and certain other seeds, and its use might possibly be extended with profit. The water should be heated only to 110° or 120° F. and the soaking may be continued until the seeds have fully swollen.

Soaking is most important with seeds having seed-cases that do not readily transmit water at growing temperatures, as in the honey locust, canna, thorn apple etc. (*27d*). Such seeds, particularly if they have been allowed to become dry, are generally soaked in hot water until swollen, before planting, otherwise they might lie in the ground for months and even years before germinating. In treating such seeds with hot water, unless the temperature at which they swell is known, the water should be heated very gradually until the seeds begin to swell, when it should be maintained at that temperature until they are fully swollen. It is said that seeds of the honey locust may be immersed for a time in boiling water without destroying their vitality, but such treatment is not to be recommended for any seeds. In seeds of this class,

37. Germination is sometimes **Hastened** by **Cracking** or **Cutting Away** part of the **Seed-Case.** To favor the absorption of water, nurserymen often drill or file a hole through the bony seed-cases of nelumbium seeds, or crack dry peach and plum pits in a vise or with an implement resembling a nutcracker (*27d*).

38. Seeds may **Fail** to **Germinate** from a variety of causes, even when exposed to the proper degree of warmth, moisture and oxygen. They may be too old (165), they may not have been sufficiently mature when gathered (163), they may have become too dry (169), they may have been subjected to freezing before sufficiently dry (167), they may have been stored while damp and thus subjected to undue heating, or they may have been damaged by insects or fungi (321) either before or after maturity. Defects of these kinds are not always visible, hence

39. Seeds should be **Tested** before **Planting** to learn if they will germinate. It is unnecessary to plant seeds in soil to test them, since the seed-tester shown in Fig. 6 is much more convenient. This useful device consists of two circular pieces of clean, moderately thick cloth of rather loose texture, a table plate that is not warped, and a pane of glass large enough to cover the plate. The cloths are dipped in water, and squeezed a few times while under the water to press out the air. They are

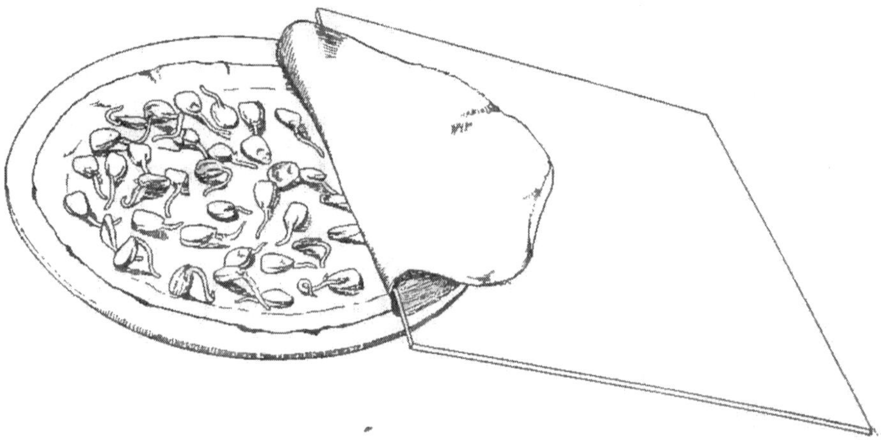

FIG. 6. Showing a simple seed-tester, adapted to farmers' and gardeners' use.

then wrung out until moderately wet, and spread over the bottom of the plate as shown, and the seeds to be tested are placed between them. It is well to use a hundred or more seeds of each sample, as a larger number will show the per cent of vitality more accurately than a smaller one, and the lot should always be well mixed before taking the sample. The plate should be kept covered with the glass to prevent evaporation from the cloths, and it may be placed in any room of comfortable living temperature. The seeds should be frequently examined, and may be removed as they sprout, when by subtracting the number that fail to sprout from the number put in, the per cent of vitality may be readily computed. The cloths should be placed in boiling water a few minutes before using them for a second test, to destroy any spores or mycelia of mold with which they may have become infected.

40. The Time Required for **Germination** varies greatly in different kinds of seeds. In lettuce seed, the tiny white shoot often breaks through the seed-case within twenty-four hours from planting, while celery seed requires several days to germinate to this extent. The seeds of many plants will not germinate the same season they are formed, even if planted under the most favorable conditions (163).

Individual seeds of the same kind and of the same sample often vary greatly in the time required for germination. Even in seeds that germinate soonest, as lettuce and radish, some individuals will not germinate until several days after the majority have germinated. Seeds of tobacco and purslane* sometimes continue to germinate through several successive seasons. The reasons for these variations are not known.

SECTION III. THE PLANTLET

By watching the germination of seeds, we may learn some interesting facts. Viable seeds will usually germinate freely on the surface of well-moistened soil or sand, if we provide a damp atmosphere above them by covering with a bell-jar or otherwise, for light does not hinder germination. One of the interesting facts connected with germination is, that the first shoot, called

* *Portulaca oleracea.*

41. The Hypocotyl* (hy′-po-co′-tyl) **Grows Downward,** on emerging from the seed-case (27*d*), no matter in what position the seed is placed. It will curve in a semicircle if necessary, to bring its rounded point in contact with the soil. But the hypocotyl is not always able to *enter* the soil, unless the seed is covered more or less, because the resistance offered by the soil is often greater than the weight of the seed. On this account, as well as to insure a supply of moisture, it is best to cover most seeds at planting, or at least to press them well into the soil (52). In nature, seeds usually become more or less covered, and those not covered generally fail to germinate.

42. The Seed-Case in **Germination.** After germination commences, the seed-case is of no further use. It has fulfilled its purpose, which is to protect the seed from the time of its maturity until the conditions arrive for germination, and is henceforth a hindrance to germination in many plants, as it must be torn asunder by the expanding plantlet. If we watch the germination of squash or pumpkin seeds through the different stages, we may discover that nature has made a special provision to help the plantlet in escaping from the seed-case in these plants. As the hypocotyl curves downward, a projection or hook is formed on the side toward the seed, which holds the seed-case down while the seed-leaves are pulled out from it. The action of this hook is shown in the accompanying figures. Sometimes, as shown in C, the point of the seed-case breaks, permitting the hook to slip off, and if the seed happens to be planted edgewise or with the point downward, the hook often fails

* Often called radicle and caulicle.

to catch the seed-case, as in D, and so the plantlet emerges from the soil without freeing itself from the seed-case and is hampered for a time. This provision is peculiar to the pumpkin family,* to which the pumpkin, squash, cucumber and melon belong, though other provisions which accomplish the same end are found in a few other families, but many plants are considerably held back by the seed-case during germination.

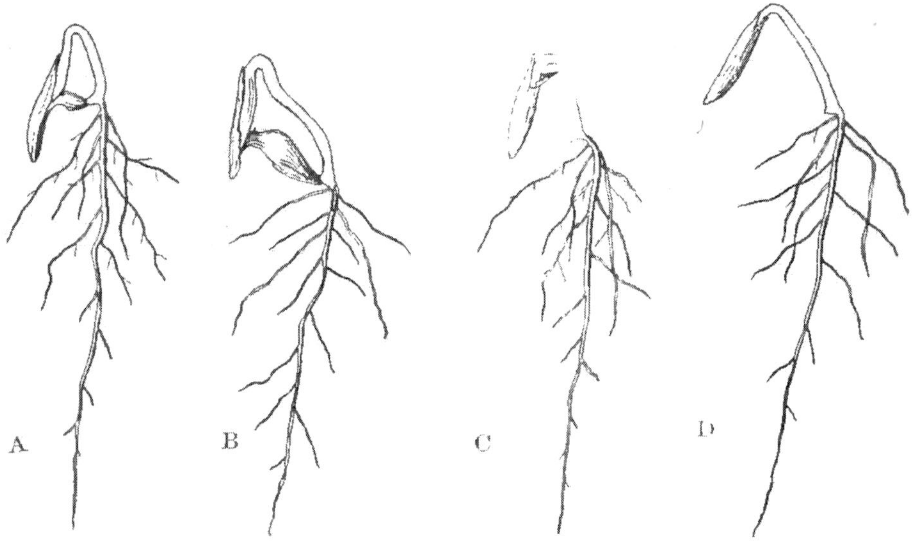

FIG. 7. Showing nature's provision to enable the pumpkin plantlet to escape from the seed-case. In A, the hook on the hypocotyl is attached to the lower half of the seed-case. B shows the same after germination is farther advanced. A fully-germinated pumpkin plantlet is shown at Fig 8.

43. Seeds of the **Pumpkin Family** should be **Planted Flatwise,** rather than edgewise or endwise, since in this position they most readily free themselves from the seed-case.

44. Some Plantlets Need Help to **Burst** the **Seed-Case.** In many seeds having hard and strong seed-cases, as the walnut, butternut and hickory nut and the pits of the

* Natural order Cucurbitaceæ.

plum, peach and cherry, the enlarging plantlet is often unable to burst the seed-case, hence germination cannot take place unless assisted by the expanding power of frost, or long exposure to moisture which softens the seed-case, or unless the seed-case is cracked before the seeds are planted (37).

45. The Roots promptly start, as the hypocotyl emerges from the seed-case — the main (*primary*) root from its point, and the branch (*lateral*) roots from its side. Sometimes root-hairs (101) may be distinctly seen, especially when seeds germinate in the seed-tester (39).

By studying Figs. 8 to 11, we may learn more of the germinating process.

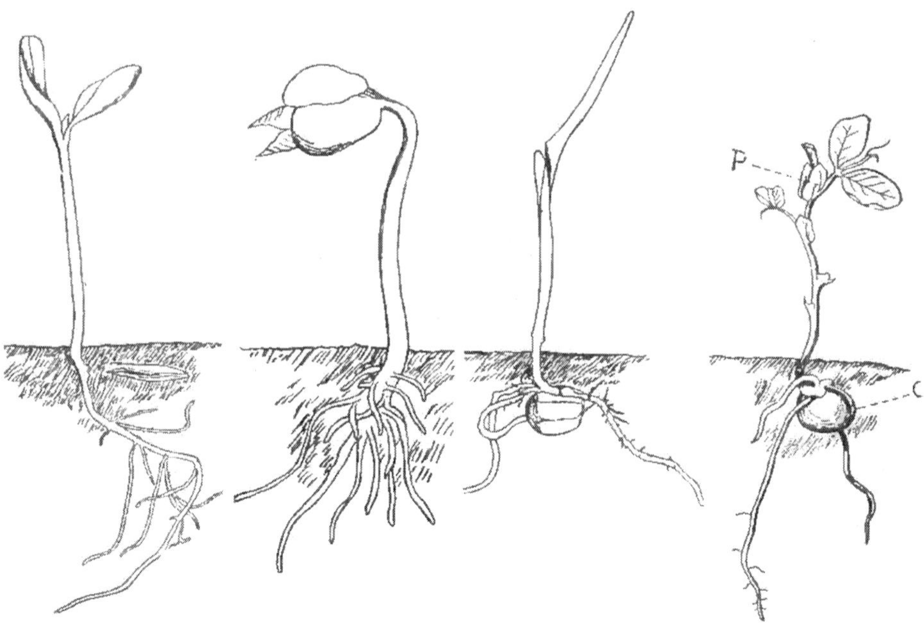

FIG. 8. Plantlet of pumpkin. FIG. 9. Plantlet of bean. FIG. 10. Plantlet of Indian corn. FIG. 11. Plantlet of pea.

In the pumpkin and bean, the seed-leaves (cotyledons) are lifted above the surface of the soil in germination.

In the pea, the cotyledons are *not* lifted above the surface of the soil in germination.

46. The Cotyledons (co-ty-le′-dons). In the bean and pumpkin, the seed, or what remains of it, seems to have separated into two parts that are united at one end — the *cotyledons* or *seed-leaves.* In the bean and pumpkin, the cotyledons form a pair of clumsy leaves, which in the bean, point downward at first, but afterwards become upright, by the straightening of the hypocotyl beneath them. We observe that the pea has also a pair of cotyledons (c), which have not separated to the same extent as those of the bean and pumpkin and are still beneath the soil. The corn, in common with other plants of its class, as sorghum, sugar cane, the reeds, grasses, etc., has but one cotyledon, and that is not easily seen without dissecting the seed. In Fig. 14, which shows a cross-section of the germinating corn grain, the cotyledon appears at *cot.*

The plants having two cotyledons form a very important class in botany, known as *Dicotyledones* (di-co-tyl-e′-dones); those having but one cotyledon form a class known as *Monocotyledones* (mo′-no-co-tyl-e′-dones). There is also a class, including the pine, fir and other conifers, that have several cotyledons.

47. The Hypocotyl Develops Differently in **Different Species.** In the pea (Fig. 11) and some other plants, the cotyledons remain in the soil, while in the bean and pumpkin, they have been lifted bodily into the air. This striking difference is due to the fact that in the pea, the hypocotyl lengthens very little in germination, while in the bean and pumpkin, it lengthens comparatively very much.

48. Seeds in which the **Hypocotyl Lengthens** in germination **Must Not** be **Deeply Planted.** When seeds of this class, which includes many plants beside the bean and pumpkin, are planted in soil, the cotyledons must be forced through the soil above them, an act requiring considerable energy. If such seeds are covered with much soil, the plantlet is often unable to lift its cotyledons to the surface, and hence must perish. Fig. 12 shows two bean plantlets that tore off their cotyledons in the vain attempt to lift them through five inches of soil. The plantlet of wheat, barley and oats, though much smaller and weaker than that of the bean, readily grows through this depth of soil, because the tiny pointed shoot (plumule (56)) of these plants readily insinuates itself between the soil particles and comes to the surface with little expenditure of energy, even when deeply planted. Plantlets of the larger beans usually fail if the seeds are planted three inches deep in a clay soil that bakes above them. Those of the castor bean,* though very robust, can hardly lift their cotyledons through one inch of soil, while those of the pea, though much more slender, readily grow through four to six inches. Apple seeds planted in autumn on clay soil, usually fail to germinate the following spring unless covered with sand or humus, or carefully mulched, because the plantlets are unable to lift their cotyledons through a baked surface soil.

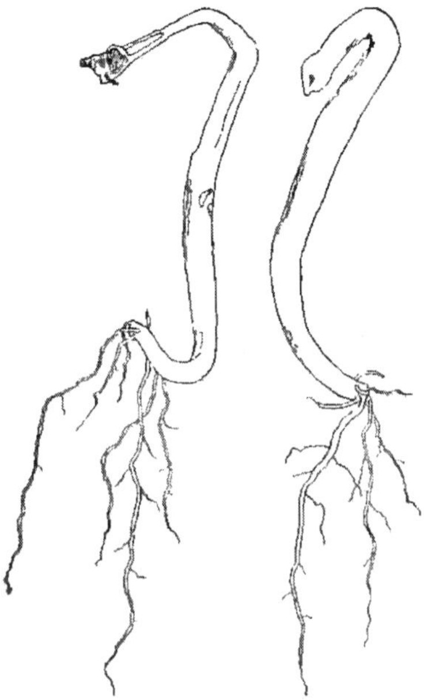

FIG. 12. Showing two bean plantlets that tore off their cotyledons from being too deeply planted.

---

* Ricinus.

49. The Vigor of the **Plantlet** is generally in **Proportion** to the **Size** of the **Seed.** This is true not only between different kinds of seeds, but between different seeds of the same kind. The larger beans, the horse chestnut and the walnut form much stronger plantlets than clover, timothy and tobacco, and the largest and plumpest specimens of any sample of seed usually form stronger plantlets than the smaller and more shrunken specimens. Growers of lettuce under glass are sometimes able to raise one more crop during the winter by sowing only the largest seeds than when the seed is sown without sifting. The practice of sifting seeds before planting, and rejecting the smaller ones, should be more generally followed (Fig. 13).

FIG. 13. Showing navy bean plants grown from large seeds (left) and from small seeds (right).

50. The Earlier Germinations from a sample of seed often **Form More Vigorous Seedlings** than the **Later Ones.** This is one of nature's methods for preserving the vigor of plants. The stronger seedlings overtop the later and feebler ones and crowd them out of existence. We should profit by this hint and reject the later plants in the seed-bed.

51. How Deep should **Seeds** be **Planted?** We have seen that one object of planting seeds in the soil is to place them in contact with moisture (29). Since the plantlet must force its way through the soil that covers the seed, the less the depth of this soil, other things equal, the less energy and the shorter time are required for the plantlet to reach the surface. Therefore, *seeds should not be planted deeper than is necessary to insure the proper supply of moisture.*

Small seeds, as of lettuce, celery and carrot, produce such weak plantlets that it is unsafe to cover them sufficiently to insure the proper moisture supply in dry weather. We must, therefore, plant such seeds so early in spring that the soil has not had time to become dry, or if necessarily planted later, we must depend largely upon artificial watering.

52. Very Small Seeds, as of petunia and tobacco, **Should Not Be Covered** with soil at all, but may be pressed down into fine loam with a board or otherwise, and must be watered often with a fine-rose watering-pot. When small seeds are sown in full exposure to sunlight, it is well to shade the surface with paper or a muslin-covered frame, to check evaporation until the plantlets appear. Small seeds are sometimes covered with a thin layer of sphagnum moss that has been rubbed through a sieve. This helps to retain moisture in the surface soil.

53. Ferns are **Grown** from **Spores*** sown on the surface of fine soil in a propagating frame (369), in which the air is kept very moist and the surface of the soil never becomes dry.

---

*Spores are the chief reproductive bodies in plants that produce no seed, as ferns, mushrooms, mosses, etc. They are usually so small as to be barely visible to the unaided eye. The dust that escapes from a puff-ball when it is squeezed or from a bunch of corn smut is formed of the spores of these plants. Spores usually consist of a single cell, in which respect they differ materially from seeds, which contain a more or less developed plantlet (54).

54. The Plantlet is Visible in the Seed. If we boil seeds of the four kinds shown in Figs. 8 to 11, or of other kinds, in water until they are fully swollen, and then carefully dissect them with the forceps and needle, using the magnifying glass when necessary, we may observe that the plantlet is present compactly folded up in the seed. Germination (28) is really little more than the unfolding and expansion of this plantlet. The plantlet as it exists in the seed is called the *embryo* (em'-bry-o).

55. The Endosperm* (en'-do-sperm). From the section of the corn grain shown in Fig. 14, it appears that in this seed, unlike the pea, bean and pumpkin, the

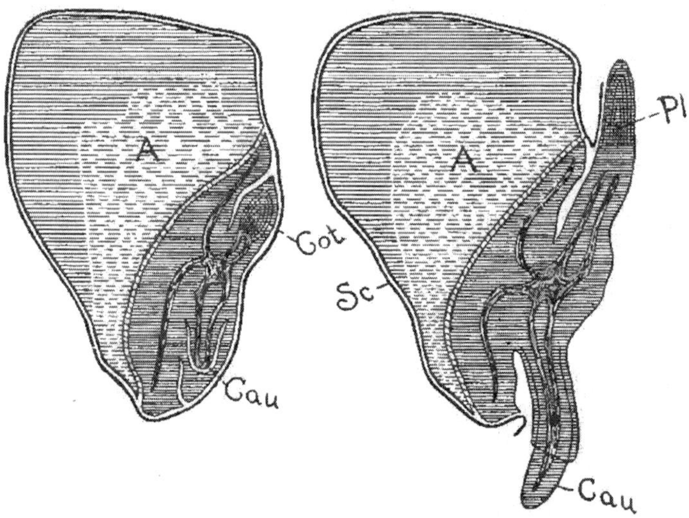

FIG. 14. Cross-section of germinating Indian corn grain. *A* endosperm; *Cot* cotyledon; *Cau* hypocotyl; *Pl* plumule. Slightly magnified. (After Frank).

plantlet and seed-case do not make up the whole bulk of the seed. The remaining part shown at A, consists mainly of cells containing starch grains and oil drops, which serve as food for the plantlet during germination,

* Called also albumen.

since active protoplasm cannot exist without nourishment (13). In the pea, bean, pumpkin and other seeds of this class, the food supply, instead of being stored by itself, as in the corn-grain, is contained within the plantlet or embryo — mainly in the fleshy cotyledons. When the food supply of the seed is separate from the embryo, as in corn and many other seeds, it is called the *endosperm.*

It is the food supply of seeds that makes them so valuable as food for animals.

56. The Plumule (plu′-mule). If we look between the cotyledons of the bean plantlet (Fig. 9), at the point of their union with the hypocotyl, we may see a pair of tiny leaves, and by carefully separating these if need be, with the point of a pin, we may discover a minute projection — the growing point (67) of the stem between them. These leaves, with the growing point, form the *plumule* — the terminal bud of the plantlet. These tiny leaves become the first true leaves, and the growing point between them develops into the stem and later leaves. By close examination, we may make out the plumule in Figs. 8, 10 and 11. In the pea and corn, it has already made considerable growth.

57. Thus we see that the plantlet or seedling consists of three parts, viz., the hypocotyl, the cotyledons (in some plants cotyledon) or seed-leaves, and the plumule or terminal bud.

58. Chlorophyll (chlo′-ro-phyll). Soon after the plantlet emerges from the seed-case, a green color appears in the parts most exposed to light. This is due to the formation within the cells of *chlorophyll* — the green coloring matter of plants. *Chlorophyll forms only in light,* and

3

when a plant containing green leaves is kept for a time in the dark, as when celery is banked up with earth, the chlorophyll disappears and the green parts become white. The chlorophyll saturates definite particles of protoplasm, called chlorophyll bodies, and since the cell-walls and protoplasm are transparent in the younger cells, the chlorophyll bodies give the parts containing

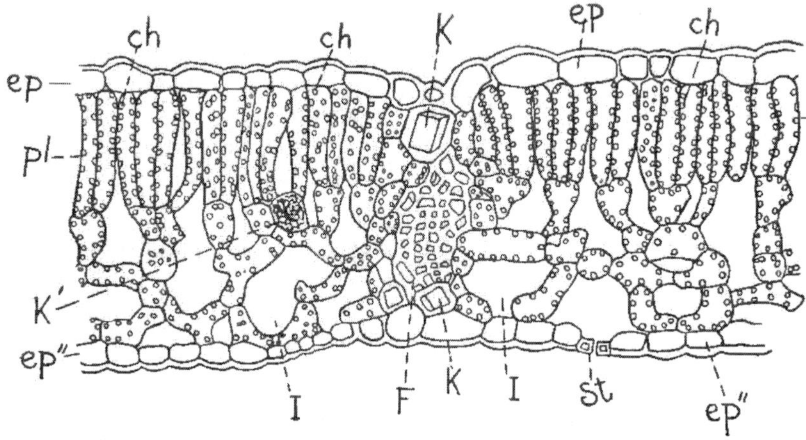

FIG. 15. Showing cross-section through leaf of Fagus sylvatica. *C* chlorophyll bodies; *Ep* epidermis of upper surface of leaf; *Ep''* epidermis of lower surface; *K* cells containing crystals; *Pl* palisade layer; *F* vascular bundle; *St* stoma; *I* spaces between the cells (intercellular spaces). Highly magnified. (After Strasburger).

them a green color. Fig. 15 shows the distribution of the chlorophyll bodies in the cells of a portion of a leaf of the beech. They appear as minute globules, which in this case are mostly located near the cell-walls. They are most numerons near the upper surface of the leaf—the part most exposed to the sun's rays.

59. No Food can be formed **Without Chlorophyll.** By the agency of chlorophyll, the chlorophyll bodies absorb energy in the form of light. This energy the chlorophyll body uses to take to pieces the carbonic acid, mineral salts and water absorbed from the air and the soil, and to recombine them into foods of various kinds which can be used by the protoplasm in making new parts and in repairing waste (*Assimilation* (as-sim′-i-la′-tion)). Until this food preparation commences, no *new* plant substance has been formed. It is true that new cell-walls and new protoplasm may be formed from the food supply of the seed before chlorophyll appears, but until chlorophyll is formed, and food preparation begins, the whole plantlet with whatever remains of the seed, when dried, weighs no more than the seed weighed at the beginning. The material formed for food is starch, or some substance of similar composition (sugar or oil), which, after undergoing chemical changes if need be, to render it soluble, is distributed throughout the plant to be built up into cell-walls and protoplasm, or to be held as reserve food (14).

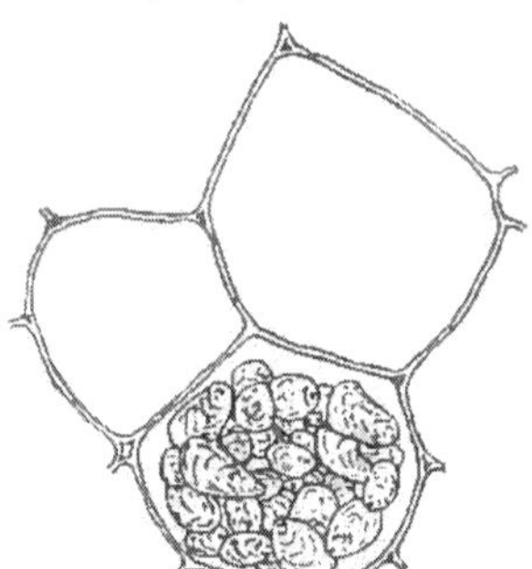

FIG. 16. Showing starch crystals stored as reserve food in a cell of potato. Highly magnified.

Food preparation and assimilation are not necessarily simultaneous, but either may proceed without the other.

Only plants can prepare food from mineral substances. The food of animals must all have been first formed by plants.

60. The Sources of Plant Food. By observing plantlets of the bean or pumpkin a few days after germination, we may discover that the cotyledons, which were at first so plump, have shriveled to a mere fraction of their former size. This change is due to the fact that the food contained by these parts has been absorbed by the developing plantlet. The patrimony furnished by the seed is quickly exhausted. Whence then comes the food that is to complete the development of the plant? Aside from the carbonic acid already mentioned (59), several other substauces are required to build up the plant structure. These are almost wholly derived from the soil, through the medium of the water absorbed by the root-hairs (101). They must all be dissolved in the soil water or they cannot enter the plant, for they must pass through the cell-walls, which are not permeable to undissolved solid matter.

61. The Elements regarded as **Essential** in the **Food** of **Plants** are carbon, hydrogen, oxygen, nitrogen, potassium, calcium, magnesium, phosphorus, iron, chlorin and sulfur. Some other elements that do not appear essential are also used by plants. All of these elements, so far as they serve as food, are absorbed by the plant in the condition of chemical compounds, as water, carbonic acid and various nitrates, sulfates, etc.

62. The Part Played by the **Different Elements.** Carbon is the chief constituent of vegetable substances and forms about half of their total dry weight. Plants obtain their carbon almost wholly from the air, in the form of carbonic acid gas, which is a compound of carbon and oxygen. The leaves absorb and decompose this gas, retaining the carbon and giving off the oxygen (59). Hydrogen and oxygen are obtained by the decomposition of water, which is a compound of hydrogen and oxygen. These enter into the construction of nearly all tissues. Nitrogen is one of the constituents of protoplasm (13). Most plants depend upon soluble nitrates in the soil for their nitrogen supply, but those of the natural order to which the clover belongs* are able to appropriate nitrogen from the air (260). Phosphorus and sulfur assist in the formation of albuminous substances; potassium assists in assimilation (59); calcium † and magnesium, while uniformly present, seem to be only incidentally useful. Iron is essential to the formation of chlorophyll (58).

Of all the materials obtained by plants from the soil, but three, aside from water, viz., nitrogen, phosphorus and potassium (254) are needed in such quantities that the plants are likely to exhaust the supply, so long as water is not deficient.

63. Water is Necessary to Growth. An adequate supply of water is the most important condition for the well-being of plants, since it not only serves in nutrition, but is the vehicle by which all other food constituents are distributed throughout the plant. Comparatively few soils are so poor as to be incapable of producing good crops when sufficiently supplied with water, while the richest soils are unproductive when inadequately supplied with it. Much of the benefit of manuring undoubtedly comes from the increased capacity it gives the soil for holding and transmitting water (93).

The supplying of food material is not the only office performed by water in the plant. The unfolding and expansion of the plantlet is largely due to a strong absorptive power for water possessed by the protoplasm within the cells. This force causes all living parts of

---

* Leguminosæ.

† Lime, which is a compound of calcium, appears to be essential to the fruiting of some plants, as the peanut, while detrimental to the fruiting of others, as the cranberry and huckleberry.

plants to be constantly saturated with water. Indeed, it distends the elastic cell-walls with water until they are like minute inflated bladders. The pressure thus set up aids in unfolding the different parts from their snug resting place within the seed-case and enables the plantlet to stand erect. Growth by cell division, it is true, begins rather early in the germination process, but this cannot take place unless the cells are first distended with water (29). A sufficient amount of water is absolutely necessary, therefore, to growth in plants. Foliage wilts in dry weather because the roots are unable to supply enough water to properly distend the cells; growth is impossible in plants of which the foliage is wilted. When the water supply is abundant, on the other hand, and the absorptive power of the roots is stimulated by a warm soil (102), the pressure within the cells often becomes sufficient to force water from the edges and tips of leaves. The drops of water that so often sparkle on foliage in the sunlight of summer mornings, commonly mistaken for dew, are usually excreted from the leaves. In young plants of the caladium, water is sometimes ejected from the leaf-tips with considerable force.

The water of plants is almost wholly absorbed by the root-hairs (101), the leaves having no power to take up water, even in wet weather. The water of plants, with its dissolved constituents, is commonly called sap, except in fruits, where it is usually called juice.

64. How Food Materials are **Distributed** through the plant. If we drop a bit of aniline blue into a glass of clear water, it will not retain its form and size, but infinitely small particles will become detached and move about to all parts of the water in which it dissolves. This movement will not stop until the bit has entirely disappeared, and until every part of the water contains exactly as much of the aniline blue as every other part. This equal distribution of the soluble material takes place in response to the *law of diffusion*, that tends to cause any soluble substance to become equally distributed throughout the liquid in which it is placed. The liquid in the meantime may remain stationary. The process would be the same if we were to put in a very small quantity of each of several soluble substances at the same time. The movements of one of these substances would not interfere much with those of the others.

If we could remove some of the dissolved aniline blue from the water in one part of the glass, it would follow that the dissolved aniline blue would move from the other parts toward this point, and if this removal were continuous, slow currents would move in this direction from all other parts of the glass.

We may now understand how the materials from which the plant is built up are distributed to its different parts. The water absorbed by the root-hairs (101) is not chemically pure, but holds in solution small quantities of various soluble matters contained by the soil, some of which are used by the plant in growth. As these useful matters are removed from the water of the cells, to be formed into food (59), the supply is replenished from the soil, not through any power of selection possessed by the plant, but in accordance with the law of diffusion. In like manner, the food formed by the chlorophyll (59) finds its way to the growing parts. Soluble matters not used by the plant are not taken in to the same extent as those that are needed, because their equal distribution is less disturbed.

The distribution of soluble matter in the plant is also promoted by transpiration (75).

SECTION IV. THE INNER STRUCTURE OF THE PLANTLET

Thus far, we have considered the plantlet mainly from the outside. Before going farther, it is well to learn also something of its inner structure. We have seen that all parts of the plant are made up of cells (12) and that these cells differ in form and office in the different parts. The cells of the leaf, for example, are different in shape and in the use they serve to the plant, from those of the stem, flower or fruit.

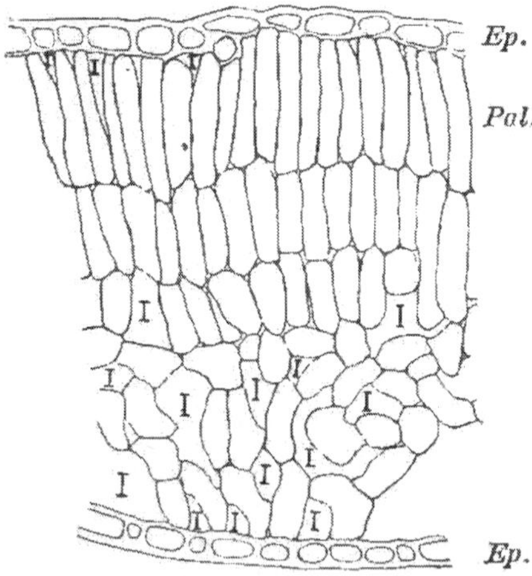

FIG. 17. Showing section through leaf of Oldenburgh apple. *Ep.* epidermis; *Pal.* palisade cells; *I* intercellular spaces. Highly magnified. See also Figs. 13 and 20.

65. The Epidermis (ep′-i-der′-mis). The plant is covered by a thin, translucent skin that extends over the entire surface of the leaves, stem and root, called the epidermis (Fig. 17 *Ep.*). This skin is formed of comparatively thick-walled cells and serves to protect the more delicate parts within. It may be readily withdrawn in some plants, as from the leaves of the liveforever * and echeveria † and young stems of the plum. The exposed surface of the epidermis of the leaves, fruit and young stems of many plants is trans-

* *Sedum telephium.* † *Cotyledon.*

formed into a layer that is more or less impervious to water, called the *cuticle* (cu'-ti-cle), which serves to restrict evaporation (75). To further protect the parts, a layer of wax (bloom) is sometimes secreted upon the outside of the cuticle, as in the fruit of many varieties of the plum and grape.

Root-hairs (101) and the hairs and bristles on the stems and leaves of many plants are cells of the epidermis elongated outward. The epidermis must not be confounded with the bark. It is replaced by bark in the older stems of woody perennial plants.

To give further strength to the upper surface of the leaf, the first two or three tiers of cells beneath the epidermis on the upper side are usually placed endwise, (*palisade cells*, Figs. 17, 15 and 3). The hardier varieties of apple, as the Oldenburgh (Duchess), have more numerous and more crowded palisade cells than less hardy varieties. Compare the palisade cells of a leaf of the Oldenburgh apple (Fig. 17), with those of Fig. 3, which shows a section from a leaf of a tender variety of apple.

66. Stomata (stom'-a-ta). Minute openings through the epidermis occur in the leaves and young stems of land plants, connecting intercellular spaces (I, Fig. 17) with the external air. These openings are each bounded by a pair of crescent-shaped guard-cells, called *stomata*, (singular, *stoma*, (sto'-ma), (Figs. 18 and 19, St). They are chiefly found on the lower side of leaves, and are extremely numerous, but are too small to be seen without the microscope. An average apple leaf has been computed to contain about 150,000 stomata to the square inch on its lower surface.

The guard-cells are delicately-balanced valves which are extremely sensitive to external influences. They are open in strong light, but usually closed in darkness and when the leaves are wet. The water escaping from leaves (75), and the carbonic acid entering them (62) mostly pass through the stomata.

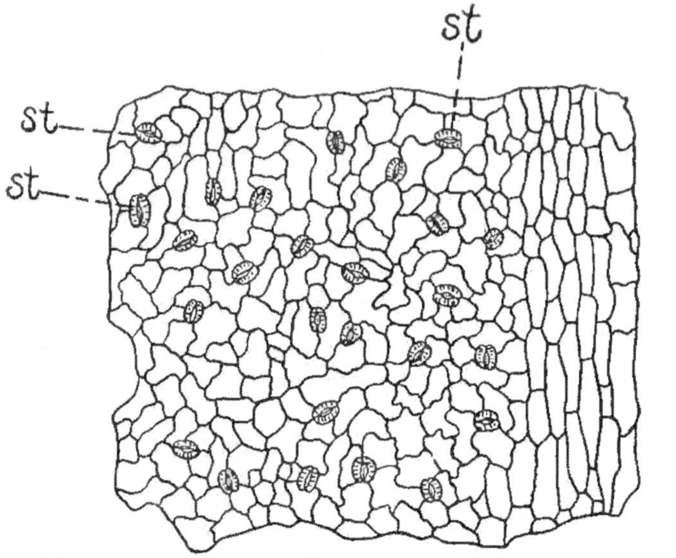

FIG. 18. Showing stomata (*st.*) on leaf of the garden beet. Moderately magnified. (After Fank and Tschirch). See also Figs. 15, 19 and 22.

The slightly-raised spots or dots on the smooth bark of the young shoots of many woody plants, (*lenticels* (len′-ti-cels)), serve a similar purpose to the stomata.

67. The Growing Point. At the tip of the stem and just behind the tip of the root, is a group of cells forming the so-called *growing point.* These cells divide very rapidly during the growing season, and from them all other kinds of cells are evolved.

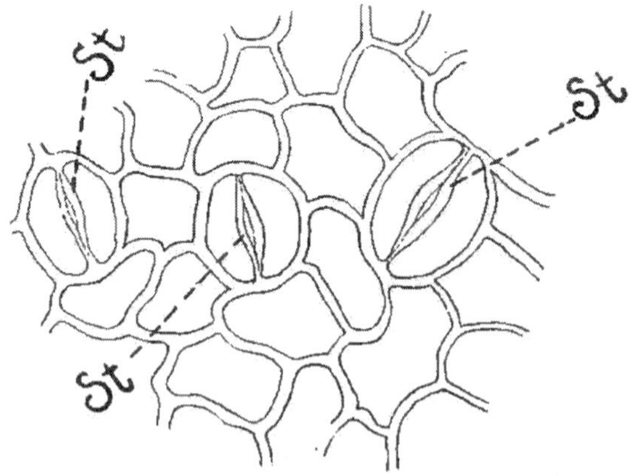

FIG. 19. Showing stomato (*st.*) on leaf of Oldenburgh apple. Highly magnified.

68. The Vascular (vas′-cu-lar) **Bundles.*** While the plantlet remains within the seed-case, it consists largely of cells more or less cubical or globular in outline. But germination scarcely commences before some of the cells begin to increase greatly in length without a corresponding increase in thickness.† These elongated cells form in groups or bundles (*vascular bundles*) that extend lengthwise through the stem and roots, and since the individual cells overlap and are in intimate contact, they form threads or fibres. These fibres serve the double purpose of giving strength to the plant and conducting water, with its dissolved food materials, to the different parts. By the absorption of the ends of some of the cells, tubes (ducts) of very considerable length are formed. In other cells of vascular bundles, the walls are much thickened and strengthened by woody deposits. These groups or bundles of fibres and ducts divide and subdivide in the leaves, forming the so-called veins and veinlets. In the roots they divide in a similar manner, extending lengthwise through all the branches and branchlets.

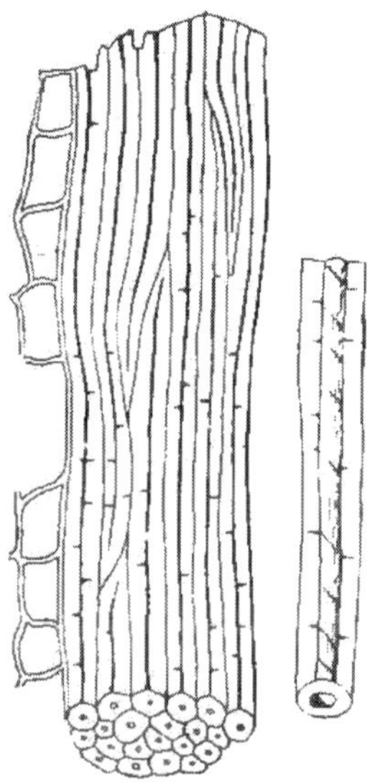

FIG. 20. Prosenchyma cells from stem of rye. Highly magnified. (After Frank and Tschirch).

Fig. 21 shows a cross-section of a vascular bundle of the sunflower.

The threads in the stalk of Indian corn and the leaf-

* Also called *fibro-vascular* bundles.

† Cells of the former class are called *Parenchyma* (pa-ren′-chy-ma), and those of the latter class *prosenchyma* (pro-sen′-chy-ma) (Fig. 20). Fig. 17 shows parenchyma cells from the apple leaf.

stem of the plantain* furnish examples of well-defined vascular bundles; in most stems the vascular bundles are less clearly defined. In woody stems they are closely crowded, which gives the wood its firm texture. In some woody plants, as the grape and the elder† a cylinder extending through the center of the stem is free from vascular bundles, forming the *pith*. The young stems of asparagus, the ball of the kohl-rabi and the roots of turnip are "stringy" when the cells of their vascular bundles become thickened by the deposit of woody material in them.

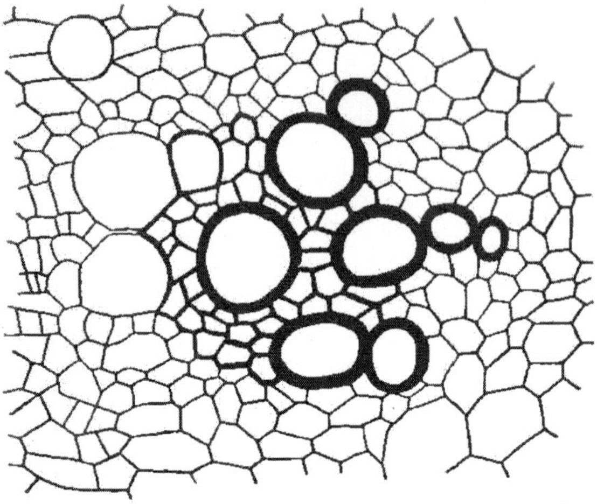

FIG. 21. Showing cross-section of a vascular bundle of the sunflower, (Helianthus annuus). Highly magnified. (After Prantl). See also Fig. 22.

69. The Cambium (cam′-bi-um) **Layer.** In most plants having two or more cotyledons (46), a layer of cells in a state of division (15) exists between the bark and the wood, called the *cambium* or *cambium layer* (Fig. 22). It is in this layer that growth *in diameter* of the stem occurs (71). The bark of plants having the cambium layer separates readily from the wood at times when growth is rapid, because the walls of the newly-formed cambium cells are extremely thin and tender. The slimy surface of growing wood, whence the bark has just been removed, is due to the protoplasm from the ruptured cam-

* Plantago. † Sambucus.

bium cells. In plants having more than one cotyledon, the cambium line is usually readily discerned in cross-sections of the stem — though it is rather more distinct and the bark is more readily separable in woody than in herbaceous* stems. In the latter, the part within the

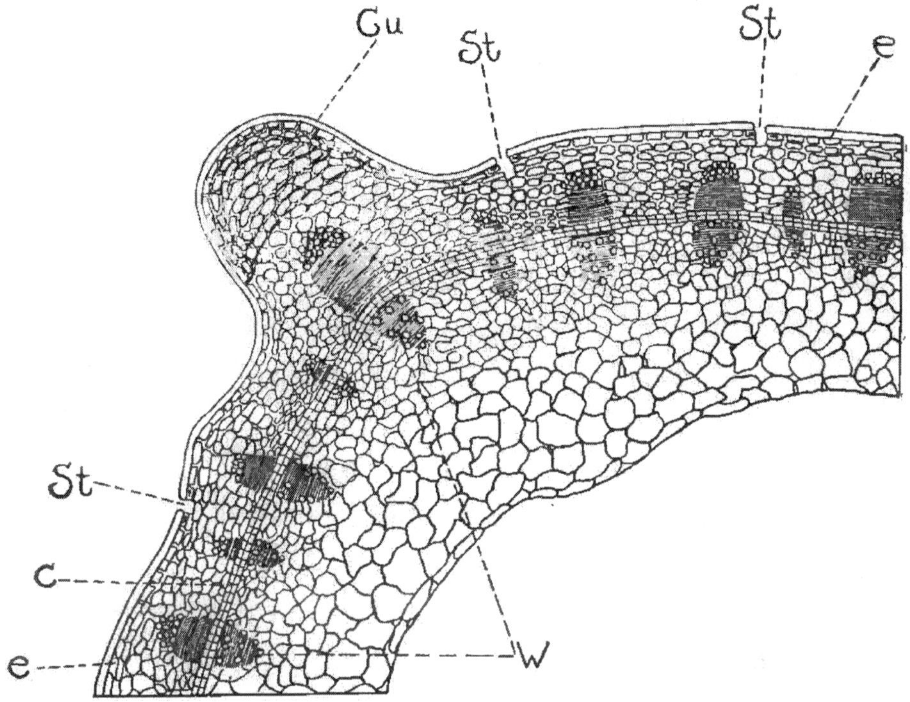

FIG. 22. Showing transverse section of corner of a bean stem (Vicia faba). *C* cambium layer; *e* epidermis; *Cu* cuticle; *St* stoma. The dark, oval-shaped spots, extending both sides of the cambium layer are the vascular bundles; *W* wood cells of the vascular bundles. Moderately magnified. (After Potter).

cambium line corresponds to the wood of woody stems, and that outside of it corresponds to the bark.

70. Portions of **Cambium** from different plants may **Unite** by **Growth.** If a section of cambium from one part of a plant is closely applied to the cambium of another part of the same plant or of another closely-related plant,

* Herbaceous stems are those that do not have the hard, firm texture of wood, as of the potato, rhubarb etc.

the two portions of cambium may unite by growth, a fact of great importance in horticulture since it renders *grafting* possible (383). Plants having no cambium layer (71) cannot be grafted, because their stems have no layer of dividing cells—the only cells that unite by growth.

71. How Stems Increase in Diameter. There is no cambium layer in plants having but one cotyledon (46), of which Indian corn, the grasses and the palms are examples. In such plants there is no clear separation between bark and wood; the stem enlarges for a time by growth throughout its whole diameter, after which it ceases to expand.

In plants having two or more cotyledons, however, additions to the bark cells are constantly being made during the growing season on the outside of the cambium layer, as are additions to the wood cells on the inside of it (Fig. 22). It follows that growth of the bark takes place on its inner surface and growth of the wood takes place on its outer surface. This explains the vertically-furrowed appearance of the bark of old trees which is being constantly split during the growing season by the forming layer within. It also explains the ringed appearance of a cross-section of a woody stem. A new ring of wood is formed each season on the outside of that previously formed, and the line separating the rings marks the point where growth in autumn ceased and was renewed the following spring. The age of a given part of the stem of a woody plant is approximately indicated by the number of its wood rings.*

* More than one wood ring is sometimes formed in a season. If growth ceases during the summer from severe drought or other cause, and is renewed the same season, an extra ring is formed.

72. The Vital Part of **Woody Stems** in plants having more than one cotyledon (46) is limited to a rather thin layer of bark and wood, of which the cambium (69) forms the center. The cells of the so-called heart-wood and those of the dry and furrowed outer bark, have lost their protoplasm, and hence are no longer alive, though they serve a useful purpose in adding strength and protection to the vital layer. The heart-wood of a tree may largely decay without materially interfering with the vital processes (Fig. 23).

FIG. 23. Live poplar tree with hollow trunk, showing to what extent the heart-wood may decay without destroying the life of a tree.

73. The Healing of **Wounds.** Cambium cells exposed to the air by partial or complete removal of the bark, soon perish, as a rule, hence growth ceases in a part of the stem thus injured. The uninjured cambium cells on the borders of the wound may, however, by division (15), form a cushion of new material that gradually extends over the injured part. A new cambium layer may thus be formed over the wound if it be not too large, so that growth of the stem may be resumed at this place. The same process occurs when a brauch is cut off near its union with the stem. The wound, if not too large, is "healed" by new growth from the adjacent, uninjured cambium cells (Fig. 24). In planted cuttings, the uninjured cambium cells at the base form the *callus* (cal'-lus) by continued division. (Fig. 25).

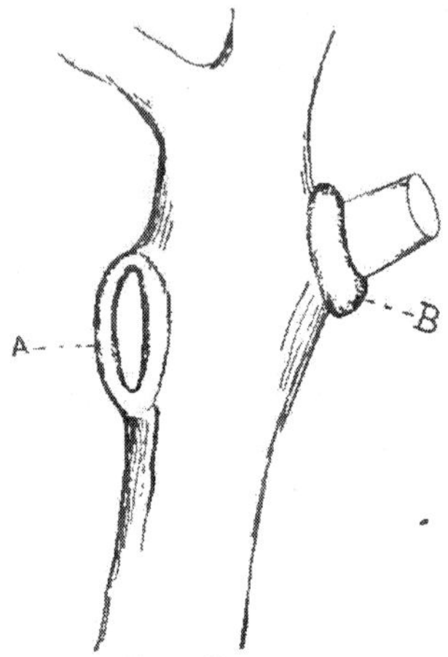

FIG. 24. Healing of wound formed by cutting off a branch (A).

FIG. 25. Showing callus at base of willow cutting.

Exposure of the bark to undue heat or cold may destroy the cambium, causing *sunscald* (186).

In periods of very rapid growth, when the cambium cells are unusually active, large areas of bark, even extending clear around the stem and as deep as the cambium layer, may sometimes be removed from trees without destroying their life, provided the recently-formed wood layer is not injured (71). In this case, the outer cells of the thin layer of cambium that remains on the surface of the wood promptly change to bark cells, hence a new bark layer forms over the exposed surface the same season.

Several successive crops of bark are sometimes removed from the trunk of the cork oak,* but in this case, the cambium layer is usually not injured.

* *Quercus suber.*

SECTION V. THE WATER OF PLANTS AND ITS MOVEMENTS

74. Plants Contain Large Amounts of Water. We have seen that the cell-walls of living plants are constantly saturated with water (63), and that the cells of the growing parts are always more or less distended with it. The proportion of water contained in living plants is generally very large. In the root of the turnip and in some fruits, it may exceed ninety per cent of the whole weight. It is greatest in young plants and in the younger and growing parts of older plants. The proportion of water is not constant in the same plants, but varies somewhat with the water content of the soil and with meteorological conditions.

75. Transpiration (trans-pi-ra′-tion). The water of plants passes off more or less rapidly from parts exposed to the air — usually as an invisible vapor. This invisible escape of water from plants is called *transpiration*. It is mainly due to evaporation of water from the plant, the same as takes place from other moist material. But fluctuations occur in the amount of transpiration from living plants that do not occur in dead organic material under similar conditions. For example, transpiration is more rapid in light than in darkness, because the stomata (66) are open in the light and thus facilitate the escape of water from the intercellular spaces. Plants poorly supplied with nourishment transpire more freely under the same conditions than those well supplied. The amount of transpiration varies greatly in different plants and depends upon the leaf surface, the nature of the epidermis and cuticle (65), the number of stomata (66) etc. Some plants, as purslane, the sedums, cacti etc., have special water-storing tissue, from which transpiration is extremely slow.

Experiments indicate that the transpiration from most leaves is between one-third and one-sixth as much as the evaporation from an equal area of water. When we take into account the immense leaf surface of a large tree, it is evident that the aggregate transpiration must be very great, as is often illustrated by the dwarfing influence of trees upon adjacent crops in dry weather (Fig. 26). Transpiration is much more rapid during dry

FIG. 26. Showing how a spruce hedge dwarfs an adjacent corn crop in dry weather.

than during wet weather, and in the rare atmosphere of high altitudes than in the denser atmosphere of low lands.

Excessive transpiration, as occurs in very dry weather, is detrimental to plants, since it reduces the water pressure within the cells below the point where healthful growth can take place (63); but normal transpiration, i. e., not sufficient in amount to interfere with healthful growth, is doubtless beneficial, since it aids in carrying food materials from the soil into the leaves (59). For this reason, plants native to regions having a rather dry atmosphere, do not thrive in greenhouses unless abundant ventilation is given to encourage transpiration.

76. Trees are **Detrimental** to **Crops** in their vicinity not only by the shade they cause, but also by their ex-exhausting effect upon the soil moisture in dry weather. The area affected by a group of trees is often much larger than is supposed. The illustration on page 58 (Fig. 26) shows how an evergreen hedge may restrict the growth of corn in an adjoining field. We should not infer from this, however, that trees are on the whole detrimental to agriculture. They serve many useful purposes.

Experimental crops intended to be comparable with eaeh other should not be planted near growing trees.

77. The Brittleness of **Young Plant Tissues** depends upon the degree of water pressure within the cells. Foliage is usually most brittle during the morning and least brittle during the latter part of the day, because transpiration is most active during the warm hours of the day. Lettuce and other salad plants are, therefore, apt to be most crisp and tender when cut in the morning. Tobacco, in which breaking of the leaves is detrimental, is preferably cut in the afternoon. Young hoed crops are generally less injured by the smoothing harrow in the afternoon than in the morning, and grass intended for hay usually dries soonest when cut in the afternoon. Lawn grass generally cuts easier in the morning than in the afternoon.

Slightly withered vegetables may have their erispness partially restored by soaking them in water for a time.

78. The Evaporation Current. Since the water of plants is taken in from the soil through the root-hairs (101), and escapes more or less rapidly by transpiration (75), it is clear that in leafy plants a current of water must pass from the roots through the stem and branches into the leaves, and that the rate of this current will depend much upon the rate of transpiration from the foliage. When the soil moisture is reduced and transpiration is excessive, this upward current of water is not always sufficient to maintain the normal pressure within the cells (63), hence the foliage wilts, or the leaves roll up, as in Indian corn and some other plants of the grass family. This current passes chiefly through the younger vascular bundles (68), which in trees constitute the so-called sap-wood, since the cells of these are less obstructed by woody deposits than those of other tissues.

The physical forces that cause the soil water to rise to the tops of the tallest trees are not well understood, but osmosis* and the pull produced by the evaporation of water from the leaves, play important parts.

79. The Flow of Sap in Spring. In the temperate zones, evaporation from the leafless stems of deciduous trees and shrubs nearly ceases during winter. The portion of the roots of these plants, however, that lies below the frost line, continues to absorb water, which gradually accumulates in the stems and branches. On the return of spring weather, the rise in temperature causes

* Osmosis is the tendency that causes two liquids of different densities to mix with each other when separated by a permeable membrane. The less dense liquid tends to flow into the denser one with a force corresponding to the difference in their densities. Cell contents are denser than soil water, hence the latter tends to flow into the cells, and thus to rise in the plant.

expansion of the tissues of the stem, as well as of the air and water within it. This creates so much pressure in some trees and shrubs that water flows freely from wounds in the wood, bearing with it, of course, the materials it holds in solution. This happens when we tap a sugar maple tree in spring. Alternate rise and fall of temperature increase the flow of sap, because with each contraction, new supplies of water or air are drawn into the stem, and thus the pressure is maintained. Sap ceases to flow on the opening of the buds, because transpiration from the foliage (75) quickly relieves the abnormal pressure.

The popular idea, that the flow of sap in spring is due to a rapid rise of water through the stem at that season, is erroneous. The sap is really rising through the stem much faster in midsummer than in early spring.

80. The Current of Prepared Food. The food of the protoplasm in the different parts of the plant is prepared almost wholly in the leaves (121). We know, however, that growth occurs in the stem and roots as well as in the leaves. It is clear, therefore, that when the stem and roots are growing, a movement of food matter must occur from the leaves into these organs. This movement may be demonstrated by a simple experiment. If a notch deep enough to pass through the bark and a little into the wood, is cut into the stem of any of our common woody plants during spring or summer, a *callus* or cushion of new cells (73) will soon form on the upper side of the notch, but not on the lower, showing that the material from which new cells are formed is passing downward. Close examination will show that this callus forms just outside the union of the bark and wood. In all plants having more than one cotyledon (46), this current is through the inner layers of the bark. The prepared food matter is dissolved in the water that saturates the cell-walls, and passes from the leaves to other parts of the plant by diffusion (64).

81. Killing Trees by **Girdling.** To destroy the life of a tree that can not be conveniently removed, we *girdle* it by cutting a notch about the trunk beneath the lowest branch. This cuts off the downward food current and so starves the protoplasm of the roots. If the notch is made after the leaves have expanded in spring, and extends only through the bark, the leaves may remain fresh for several weeks, for the transpiration current passing through the sap-wood (78) may continue. Since the roots receive no nourishment however, they will soon cease to grow and will usually die from starvation before the following spring. If the notch is cut deep enough to reach through the sap-wood, thus cutting off both the ascending and descending currents, death of the tree soon follows.

82. Root Starvation may occur **Without Girdling.** In seasons of extreme drought, when the leaves are poorly snpplied with crude food materials from the soil, the amount of prepared food may be so meagre that the food current will be exhausted before it reaches the roots. In such cases the roots perish, and the tree is found dead the following spring. This most frequently occurs with trees on poor soil, that have suffered from insect attacks as well as from a dearth of water. It often occurs also in recently-transplanted trees that fail to make satisfactory growth the first season.

83. To Destroy the most persistent **Weeds** we starve the roots by preventing all leaf growth (339).

84. Restriction of the **Growth Current Promotes Fruitfulness** by causing an accumulation of prepared food in the stem and branches (135 B).

85. The Storage of **Reserve Food.** In healthy plants, food is usually prepared faster than it is consumed by growth. The surplus may be in the form of starch, as in the potato (Fig. 16), wheat and sago; sugar, as in the sugar cane, sugar maple and beet; or oil, as is cotton seed, flax seed and rape. Aside from the seeds, which are always stocked with reserve food, certain plants living more than one year, as the potato, beet, onion, kohlrabi etc., have special accumulations of food in certain parts, and the parts of plants that contain such reserve food are most valuable as food for man or animals. The proportion of starch stored in potato tubers is not constant, hence the food value of different samples of potatoes may vary greatly. In woody plants, the surplus food is more evenly distributed through the different parts, though the older leaf-bearing wood is usually best supplied.

86. Plants Use their **Reserve Food** in the production of flowers and seeds (135 A), and in repairing damages, as the healing of wounds (73), or the replacement of leaves destroyed by insects or otherwise. Annual plants (337) expend all their reserve food in flower and seed production and then perish as soon as the seed is ripe. Biennial plants devote the first season of their life to storing an abundant food supply, which is expended in flower and seed production the second year. Our seed crops, as oats, corn, peas and beans, are mostly annuals; our vegetables other than seeds, as beets, cabbage, parsnips and celery, are mostly biennials. Perennial plants, in normal condition, expend only a part of their reserve food in any one season for the production of flowers and seeds, withholding the remainder for nourishment through the winter and to develop leaves the following spring. The reserve food in dormant cuttings (358) enables them to form roots and expand their buds.

SECTION VI. THE ROOT AND THE SOIL

With the out-door cultivator, the part of the plant environment that lies beneath the soil surface is more under control than the part that lies above it. He can do little to change the composition or temperature of the air or the amount of sunlight; he may do much to influence the fertility, the texture, the drainage and the aeration of the soil. A knowledge of the roots of plants and of the soil in which they grow and feed, is therefore, of the utmost practical importance.

87. The Office of the **Root.** The roots of land plants serve (a) to anchor the plant in the soil, enabling the stem or stems of erect species to grow upright, and (b) to supply the plant with water with its dissolved food materials (63).

88. The Root Originates in the **Stem.** As we have seen the primary root develops from the lower or "root-end" of the hypocotyl (45). But lateral roots may develop freely from other parts of the stem. If we examine the base of the stem of a plant of Indian corn a few weeks after planting, we may see that the main roots start above the point at which the stem was originally attached to the seed; and if we pull up a pumpkin vine or an untrellised tomato plant late in summer, we often find it rooted from the stem at some distance from the original root. Lateral roots originate in the internal tissues of the stem or root and not close to the surface, as do buds (132).

89. Moisture Excites Root Growth. Roots develop, as a rule, from portions of the stem that are maintained for a certain time in contact with abundant moisture. In the pumpkin vine and tomato plant above mentioned, nearness to the soil furnishes a moist atmosphere. A corn-stalk pegged down to the ground for some distance will usually root at all joints of the stem in contact with the soil. A potato plant grown under a bell-jar, where the air is nearly saturated with water, will form roots at any joint of the stem. In parts of the tropics where the air is very moist, certain plants, as orchids and the Banyan tree,* emit roots freely from the stem above ground. Cuttings (358) and layers (349) form roots because they are maintained in contact with abundant moisture and at a suitable temperature. Cuttings of some plants, as the willow and nasturtium,† root promptly when their stems are immersed in water.

90. Oxygen is Necessary to the Life of Roots. Since the cells of newly-formed roots are filled with protoplasm, they must have access to the oxygen of the air, or they can neither live nor grow. This is shown by a simple experiment. Boil a quantity of water fifteen

---

* *Ficus Indica.* † *Tropæolum.*

minutes or longer, to exhaust it of free oxygen, and then cool it quickly by setting the dish containing it in cold water. Now place a healthy cutting (358) of some plant that roots freely in water, as willow, nasturtium or wandering jew,* in each of two tumblers. Pour a part of the cool, boiled water into one of the tumblers and add a little olive oil to form a film over the liquid thus preventing it from absorbing more air. Then agitate the rest of the water vigorously to impregnate it again with oxygen, and pour some of this into the second tumbler. Set both tumblers in a light, warm place. In a few days roots will start freely from the slip in the tumbler in which the water has access to the air, but not in the other (Fig. 27). If now the rooted cutting is placed in oil-covered water that has been exhausted of its oxygen by boiling, the roots will soon die.

FIG. 27. Slips of Tradescantia in water containing oxygen (left glass) and in water containing no oxygen (right glass). From nature.

The copious formation of root-hairs (101) that reach out into the moist atmosphere of the seed-tester (39), and that so often fills the soil cavities with a delicate, cottony down, is

* *Tradescantia.*

further proof that roots search for air as well as water. The total absence of live rootlets in the puddled clods of badly-tilled fields shows that roots will not penetrate soil from which the air has been expelled by undue compression while wet. Plants in over-watered greenhouse pots sometimes send rootlets into the air above the soil to secure the oxygen from which their roots have been deprived.

91. The Ideal Soil for **Land Plants** must contain enough plant food and water to fully supply the plants, and yet be so porous that air can circulate through it and come in contact with the roots. Each particle of such a soil is surrounded by a thin film of water, while between the particles are spaces connected with each other, and filled with moist air that is in communication with the air above the soil. The root-hairs (101) apply themselves intimately to the wet surfaces of the soil particles, or reach out into cavities filled with saturated air, and are thus able to draw in the well-aerated soil water, with its dissolved food constituents, in sufficient quantity to restore the loss from transpiration (75) and to distend the newly-formed cells (63).

92. The Soil is a **Scene** of **Constant Changes.** The part of the soil in which the roots of plants grow is the field of most potent vital and chemical activities. The dead remains of plants and animals it chances to contain are undergoing decomposition during the warm season, by serving as the feeding ground of myriads of microscopic plants (bacteria). Through their agency nitric acid, which supplies the higher plants with their most valuable food element — nitrogen (255), is formed in the soil.

The carbonic acid these remains took from the air during growth is also set free to slowly disintegrate the mineral soil constituents, rendering these soluble and thus available as plant food. In winter, the frost separates the compacted particles of clods, making the latter permeable to air and rootlets, or flakes off new fragments of rock, thus unlocking new supplies of mineral fertility.

93. The Importance of Organic Matter in the Soil. Crops secure a large part of their nitrogen, as well as of other food substances, from dead organic matter, i. e., animal or vegetable materials. The application of such matter to the soil is, therefore, of great importance, where large crops are expected. Stable and barn-yard manure, the offal from slaughter-houses, tanneries, breweries etc., are all valuable for this purpose, when wisely used. Not only does organic matter in the soil furnish plant food, but while in a partially decomposed state (humus), it renders the soil porous and greatly increases its water-holding power.

94. The Soil Needs Ventilation. The roots of growing plants and the decomposition of organic matter in the soil tend constantly to exhaust the latter of its free oxygen, and to replace this with carbonic acid, which is not used by the roots. Hence, without some interchange between the contents of the soil cavities and the atmosphere above, the roots sooner or later become smothered and perish. In sufficiently porous soil, changes in temperature and in atmospheric pressure, aided by wind and rain, furnish the needed soil ventilation, but in poorly-drained soils, and soils not thoroughly tilled, the roots of plants often suffer from insufficient oxygen. A puddled crust on the surface of clayey soil, due to the compacting influence of rain, is a great hindrance to its ventilation. Earthworms and other animals that burrow in the soil aid in aerating it.

95. Hotbeds Require Especial Care in **Ventilation** (365), since they usually contain large quantities of decomposing organic matter (manure), which rapidly absorbs oxygen from the soil, replacing it with carbonic acid.

96. Drainage Promotes Soil Aeration by forming an outlet for the surplus water that would otherwise fill the cavities. Although moisture is essential to root growth, land plants do not prosper with their roots immersed in water. True, most plants may be grown in "water culture," i. e., with their roots from germination grown in water that is freely exposed to the air; but the roots of land plants soon smother for want of free oxygen when the soil cavities are filled with water, because the soil tends to prevent the water within its cavities from absorbing air.

97. Potted Plants Require Drainage (412), and the outside of the pots should be kept clean, to admit air through their walls. Potting soil should contain sufficient sand and humus (93), so that it does not readily become puddled by watering (31).

98. Potted Plants should be **Watered** with **Care** (219). They should receive sufficient water so that the soil particles are constantly surrounded with a film of water, but not so much that the soil cavities remain filled.

99. How the **Root-Tip Penetrates** the **Soil.** Darwin made the interesting discovery that the root-tip, in advancing through the soil, does not move in a straight line, but has an oscillating motion, which enables it to take advantage of openings between the soil particles. The force with which the root-tip is pushed forward was calculated by Darwin to be at least a quarter of a pound in some cases, while the increase of the root in diameter may exert a much greater force. The root-tip is protected in its passage through the soil by a thimble-like covering called the *root-cap.**

100. Growth of Roots in Length. Since the soil offers more or less resistance to the growth of roots, it is evident that the roots of land plants cannot elongate through their whole length at once. On the contrary, the part that increases in length is limited to a short portion just behind the root-tip. Sachs found that the part of the rootlet of the broad bean that increased in length by growth scarcely exceeded half an inch long. In Fig. 28, the parts that are increasing in length are considerably shorter than the root-tips (R T).

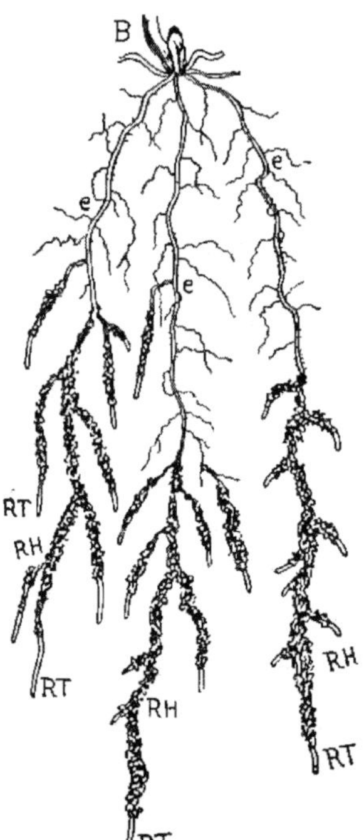

FIG. 28. Roots of young wheat plant. The parts inclosed in sand (R H) are surrounded by root-hairs. R T, root-tips; e, older parts of root. One-fourth natural size. After Frank and Tschirch.

101. The Root-Hairs (Fig. 29 B.) develop just behind the elongating part of the rootlet and are present in nearly all plants. Their object is to absorb water, with the food materials it contains. The root-hairs greatly increase the absorbing surface of the roots, just as leaves increase the absorbing surface of the plant above ground.

* The root-cap is readily seen without a magnifying glass when a bean plant is grown in water.

Each root-hair consists of a single elongated cell (Fig. 30), and is filled with protoplasm, as are the cells in other living parts of the plant (13). As the extremity of the root advances through the soil by growth, new root-hairs are formed in front of the older ones, while those farthest back as rapidly die off, so that only a short portion of a rootlet bears root-hairs at any one time. In Fig. 27 root-hairs are visible in the left glass, and in Fig. 6 they may be seen on the hypocotyl of some of the germinating corn grains. In Fig. 29A and in Fig. 28 the parts of the root bearing root-hairs are indicated by the sand which adheres to these parts. It is usually difficult to see the root-hairs of plants growing in the natural soil, but they may sometimes be discovered with the help of a pocket magnifying glass by carefully removing the soil particles about the younger roots, when the silky network of root-hairs may be seen filling the smaller pores of the soil or enveloping the soil particles. Fig. 30 shows a magnified root-hair

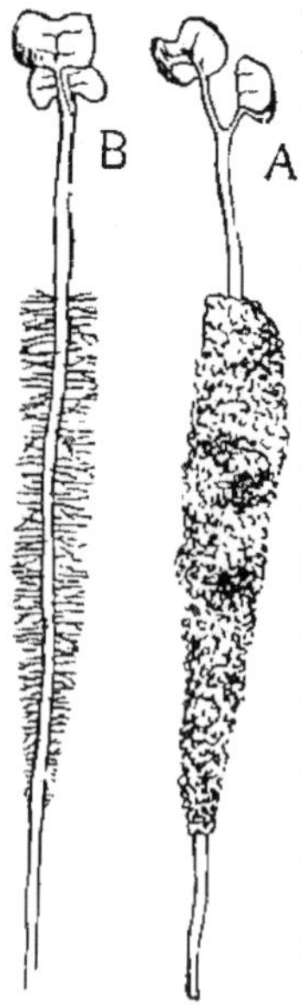

FIG. 29. Seedlings of turnip showing root-hairs. (After Frank and Tschirch).

FIG. 30. Magnified root-hair of wheat, in contact with soil particles. (After Sachs).

of the wheat plant, closely attached to some particles of soil. The root-hairs are able to take up water freely, even from soil that does not appear very wet, because each soil particle is enveloped in a thin layer of water (91). Still more interesting is the fact, that root-hairs are able to dissolve mineral matters in the soil, by their excretions most important of which is carbonic acid, thus permitting the plant to use these matters as food.

102. Root-Hairs Absorb Water with considerable force. It is the absorptive power of the root-hairs that causes water (sap) to flow so freely from injured stems of grape vines* and some other plants in spring, and from wounds in the trunks of some trees in summer. This force is probably due to the absorptive power of the protoplasm in the very active young root cells. It is affected by the temperature of the soil within certain limits, lessening as the temperature falls, and increasing as it rises. Sachs found that the foliage of plants of tobacco and pumpkin drooped when the temperature of the soil in which they were growing was reduced much below 55° F., showing that the roots did not absorb enough water at that temperature to compensate for the loss by transpiration (75). When the soil is warm, on the other hand, the absorptive power of roots may be sufficient to force water from the tips of leaves during cool nights when transpiration is slight (63).

103. Only the Youngest Parts of Roots are Active in Absorption. The part from which the root-hairs have perished absorbs little water, but is chiefly useful in giving strength to the plant and in conducting the plant fluids. The absorbing part of any given rootlet is, therefore, comparatively short. It follows that the amount of nourishment a given plant can receive will depend upon the *number* of its root-tips. Our treatment

---

* Hales found the absorbing force of the roots of a grape vine equal to the weight of a column of mercury thirty-two and one-half inches high.

of the plant should, therefore, be aimed at promoting the formation of root-tips. In other words, *we should encourage root branching.** How may we do this?

104. The Branching of **Roots** in land plants appears to depend much upon the amount of free oxygen (31) and available plant food which the soil contains, so long as the moisture supply is sufficient. In cultivated grounds having a compact sub-soil the roots of annual crops usually branch most freely just at the bottom of, or a little below, the layer of soil stirred by the plow, this being the point at which the supply of oxygen, plant food and moisture are probably best suited to root growth. As the depth of tillage is increased, roots branch freely at a greater depth. Masses of decomposed manure beneath the surface of the soil are usually penetrated through and through with finely-branched roots; and fragments of bone in the soil are often inclosed in a mat of delicate rootlets. These materials furnish plant food in abundance. Roots that penetrate the deeper and more compact layers of soil, on the other hand, and those in poor and dry soils, are usually little branched. It is clear, therefore, that unless a soil is well aerated (94) by a proper system of tillage, and by

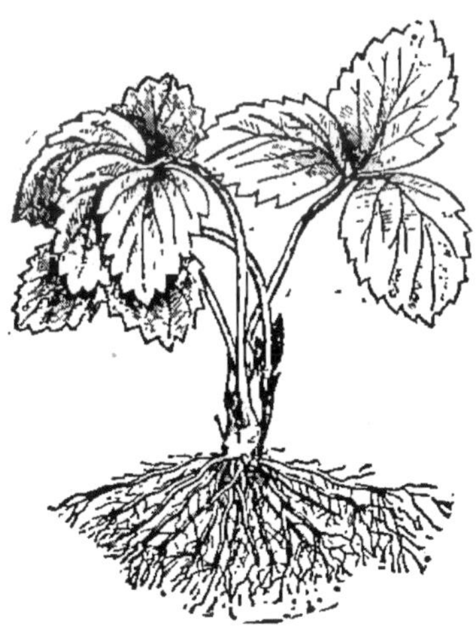

FIG. 31. Showing how root pruning stimulates root branching.

* Root branches must not be confounded with root-hairs. In Fig. 28, branches of the roots appear at e. e. e. The branches bear root-hairs when of sufficient length, but root-hairs never develop into branches.

draining if need be, and unless it contains abundant soluble plant food in the aerated part, the roots of plants growing upon it will not branch freely and hence the plants cannot be well nourished.

105. Transplanting (400) and **Root Pruning** (416) **Stimulate Root Branching.** Removing the growing points of either the stem or root (67) stimulates the development

Fig. 32. Showing effects of transplanting on root growth of celery plants. The left two plants were transplanted when quite small; the right two were not. (After Green.)

of other growing points farther back. Transplanting or root pruning accomplishes this in the case of roots (Fig. 31, p. 73). While these operations may not often increase the total number of root-tips, and hence may not enable the plant to take up a greater amount of nourishment, they do cause the development of a more compact root system, which is of great advantage to young plants grown in the seed-bed or nursery for subsequent transplanting.

106. Pricking Off Young Seedlings, i. e., transplanting them from the soil in which they grew to other soil, where they have more room, is an important preparation for their final transplanting. They should receive as good care after pricking off as before, with which they soon develop many new rootlets near the base of the stem, that need be little injured in the later removal (Fig. 32, p. 74).

107. Nursery Trees are **Benefited** by **Transplanting** them once or twice before the final planting out, for the reasons named above.

108. Root Pruning (416 k) is sometimes employed as a substitute for transplanting, and is especially useful to trees that form few branch roots, as the hickory and walnut. In this case, the tap root is cut off a few inches below the surface of the soil the year before transplanting.

109. The Horizontal Extent of **Roots** is usually greater than is generally supposed. In upright-growing plants, the area occupied by the roots, as a rule, exceeds that covered by the foliage, while in spreading and trailing plants, the roots are probably not often less in extent than the branches. It appears from the observations recorded that even in such plants as the melon and squash, the horizontal extent of the roots usually equals or exceeds that of the runners. As the diffusion of soluble matters in the soil water is probably much hindered by the soil particles, the roots of plants need to travel farther after food than do the branches, which develop in a freely circulating medium. Especially is this true of plants growing in poor soil.

110. The Depth of **Roots** in the **Soil.** It appears from the observations recorded that the extreme depth reached by roots is generally less than their greatest horizontal extent. The distance reached by the deeper roots is probably governed largely by the nature of the subsoil and the depth of free ground water. But in most annual crops a comparatively small part of the root system develops much below the plow line. At the Geneva Experiment Station * the chief root-feeding ground of the field and garden crops grown in that locality appeared to be from three to ten inches below the surface, while that of crops making large development of stem and foliage during summer, as Indian corn, sorghum, tobacco and the Cucurbitæ, appeared to be shallower than in slower-growing crops.

A portion of the roots of many crops grow very near the surface of the ground. Branches from the main horizontal roots often grow upward as well as in other directions. At the Geneva Experiment Station, numerous roots of sweet corn were found within an inch of the surface, and in a tall-growing southern corn, roots of considerable size started at a depth of only half an inch. The main root of a Hubbard squash vine was traced a distance of ten feet, in which its depth varied from two to five inches. In tobacco fields, the rootlets sometimes literally protrude from the surface of the soil in warm, wet weather (232).

III. The Rate of Root Growth in rapidly developing plants is often extremely fast. President Clark, formerly of the Massachusetts Agricultural College, concluded from very careful examinations and measurements of the roots of a squash vine grown under glass, that rootlets

* See Report of New York Agricultural Experiment Station, 1886, p. 165.

must have been produced at the rate of at least one thousand feet per day during the latter part of the growth period.

112. Relation of Roots to Food Supply. In the extent of ground occupied, root growth is relatively less in moist and fertile soils than in poorer and drier ones, but the roots are proportionally more branched. In wet seasons, a given plant has less extensive root development than in drier seasons, because the roots may then secure the needed food and water from a smaller area. Nursery trees grown on fertile soils have a more compact root system than those grown on poorer soils.

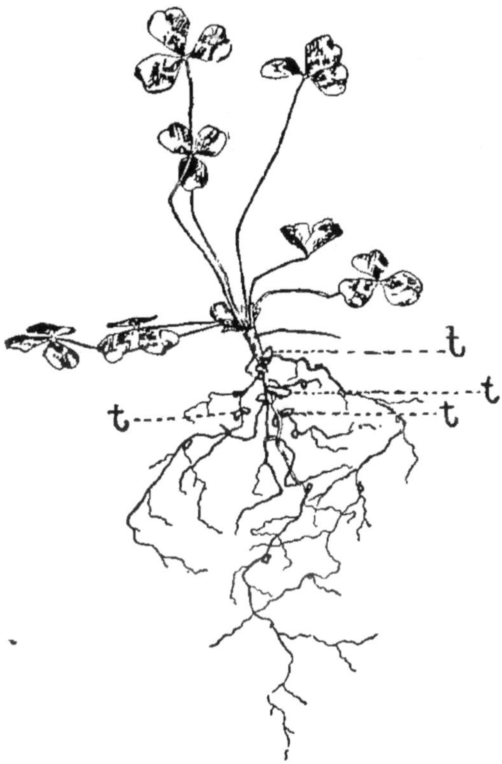

FIG. 33. Young clover plant showing tubercles on roots (t). (From nature).

113. Root Tubercles. Plants belonging to the natural order *Leguminosæ* (le-gu-mi-no′-sæ), of which the clover, pea and bean are familiar examples, when grown in ordinary soil have swellings or tubercles on their roots (Fig. 33). These are caused by micro-organisms, probably of the class known as bacteria, and are of special interest, because the organisms producing them render nitrogen of the air available as plant food. Plants have no power to utilize directly the free nitrogen of the air (260).

SECTION VII. THE STEM

114. As the root develops from the base of the hypocotyl, the plumule, or primary shoot (56), develops from the other end and becomes, at least for a time, the main axis or stem of the plant.

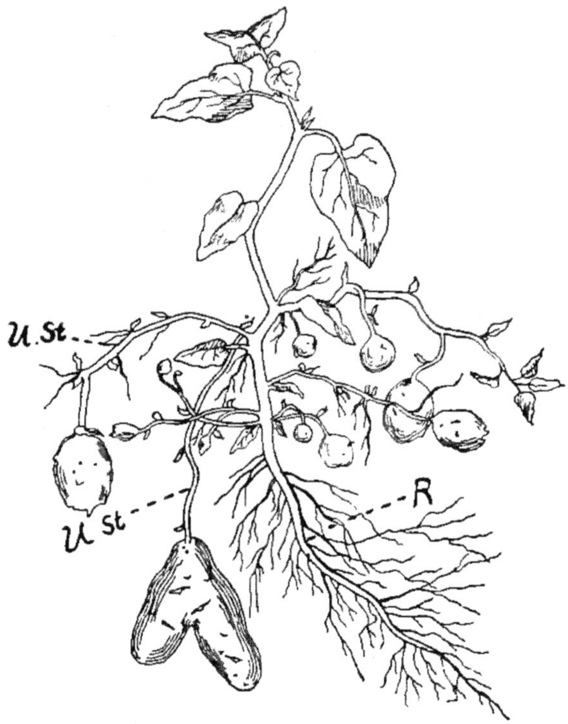

FIG. 34. Potato plant. U. st., underground stems; R, roots. The tubers are the thickened distal* ends of the underground stems. Much reduced. (After Frank and Tschirch.)

115. The Stem is, generally speaking, the part of the plant that supports the leaves. In exceptional cases, as in the potato (Fig. 34) and quack grass, a part of the stem grows beneath the ground, on which the leaves usually do not develop (*underground stems*); and in a few plants, as in some cacti, the stem performs the whole office of leaves. The stem may be strong enough to support its own weight, as in trees and shrubs, or it may depend upon other objects for its support, as in vines.

116. Nodes and Internodes. Unlike the root, the stem is developed in successive sections, comparable in part to the stories of a building. Each section or story

---

* See foot note on page 79.

consists of one or more leaves, attached to the distal* end of a portion of the stem. The part of the stem to which the leaf or leaves are attached is called a *node* and the part below the node, or in the stem as a whole, the part between the nodes, is called an *internode*.

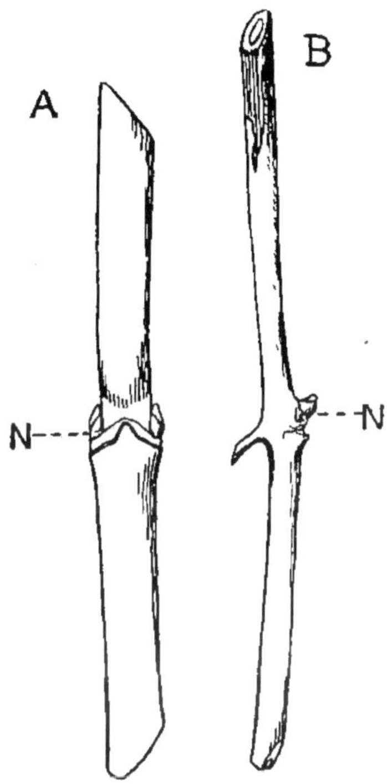

FIG. 35. Nodes (N); A, of the box elder, *Negundo aceroides*; B, of the wild grape, *Vitis riparia*.

The nodes are distinctly marked in the younger stems of most plants by a slight enlargement or by leaf-scars, if the leaves have fallen (Fig. 35). The nodes are centers of vital activity and are the points at which lateral growing points (buds (128)) are normally formed, and whence roots usually start first in cuttings and layers (358, 349).

117. The Stem Lengthens by **Elongation** of the **Internodes,** as well as by the formation of new ones. As the internodes soon attain their ultimate length, it follows that the stem lengthens only near its distal end. An internode that has once ceased elongating does not usually resume it, hence the internodes of perennial plants that are only partially elongated at the close of the growing season in general remain undeveloped. When growth is resumed in spring, the formation of a comparatively long internode beyond the very short ones of autumn usually forms a perceptible ring about the shoot, which enables us to readily

---

* *Distal* means farthest from the point of origin, i. e., the point at which growth started. It is opposed to *proximal*, which means nearest the point of origin.

locate the point at which growth started in spring (Fig. 36). Indeed we can often determine the amount of growth that took place during the preceding season or even farther back.

118. The Ultimate Length of the **Internodes** in any plant, or any part of a plant, depends upon the rate of growth—rapid growth producing long internodes, and vice versa. In the same species, therefore, the average length of the internodes is much greater in vigorous, young plants than in old ones; in the main, central shoot than in the branches, and when growth is well started in spring than during its decline in autumn. The diameter of young internodes that are not unduly shaded is generally in proportion to their length, hence rapidly-growing shoots are usually thicker than slower-growing ones. We can judge of the comparative vigor of nursery trees by observing the length and diameter of the internodes.

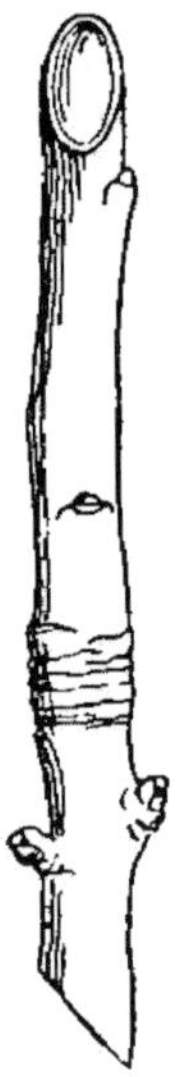

FIG. 36. Union of new and older wood

119. The Stem Elongates Fastest just behind the growing point (67), and at least in young plants, just behind the primary or original growing point (56). When we desire to check growth of the stem, therefore, we remove the terminal growing point by *pinching* (416 a).

120. Pinching Stimulates Branching because removing the terminal growing point stimulates the development of other growing points farther back (105).

SECTION VIII. THE LEAVES

We have seen that one or more leaves are normally formed at each node of the stem (116).

121. The Function of Leaves is food preparation (59). Since food is prepared only in the light, the cells of leaves are in most plants so arranged as to best expose them to light, i. e., in thin, more or less horizontal plates, which are strengthened and at the same time supplied with water by a network of vascular bundles (68) connecting with the stem. They are protected by the epidermis (65), but have access to air through the stomata (66).

Each leaf, like the stem and root, is developed from one or more growing points (67), one of which forms the terminus of each lobe or division of the leaf. Cell division in the leaf is confined to the near vicinity of the growing points, hence an injury to the older part of the leaf is not repaired further than by the formation of callus (73) over the wounded parts.

122. The Cultivator Should Provide for Normal Leaf Development. Since the protoplasm of the plant is nourished by prepared food (59), and since food preparation in most plants takes place almost wholly in the leaves (121), it is of first importance that the plant be so cared for as to promote normal leaf development. *Without this, good crops are impossible.* The plants must be grown far enough apart so as not to unduly shade each other; insects and fungi must not be permitted to prey upon them when it is possible to prevent it; and the leaves must not be needlessly removed or injured.

123. How Far Apart Should Plants be Grown? When the finest developed plants, or parts of plants, as fruits, flowers, leaves, stems or roots is desired, the plants should not be grown so near together as to interfere with each other's leaf or root development. But when the largest crop from a given area is of more importance than the development of the individual plant, as with grain crops, the loss from a limited amount of shade and crowding will be more than made up by the increased number of plants. In this case, the amount of crowding that will give the maximum yield will depend much upon the fertility and moisture of the soil, and must generally be determined by experiment.

124. Stem and Root Development Depend on the Number of Leaves. Since the vascular bundles, through the formation of which the stem and root increase in diameter, originate in the leaves (68), the size and firmness of the stem and the root depend somewhat upon the number of leaves the plant bears. The more leaves it has, the more solar energy it can transform into plant tissue. The stem is larger beneath a vigorous leafy branch, and if cut off some distance above a branch, the part thus deprived of its foliage ceases to grow, unless it develops new leaves. Trees growing in the dense forest, where their lower branches continually perish through lack of light, have tall, but very slender trunks, and their wood is soft because it contains comparatively little fibrous tissue, while other trees of the same species, in the full light of the open field, through the large amount of solar energy absorbed by an immense number of leaves, develop massive trunks, of which the wood, being packed with fibrous tissue, is much stronger than that of the forest tree.

125. The Comparative Size of Leaves on a given plant depends much on the water supply during their formation. The leaves of sap-sprouts (224), that take an undue proportion of water, are usually very large, and in upright-growing plants, the leaves on the more nearly vertical shoots are usually larger than those on the horizontal ones. The more vigorous the plant, the larger, as a rule, are its leaves, aud the softer is its woody tissue.

In plants grown from seed to secure new varieties, large leaves may be taken as evidence of superior root development, which implies capacity to endure drought and, therefore, hardiness. In the apple, the large-leafed varieties are, as a rule, hardier than others, probably because their vigorous roots supply the needed water during the dry season, thus enabling the tree to mature healthy wood and buds which can pass severe winters unharmed (175).

Crops grown for their leaves, as cabbage, lettuce, tobacco etc., are especially liable to be curtailed by drought, and hence should be given the culture that best promotes soil moisture, as abundant surface tillage and liberal manuring (232).

126. Leaves are usually **Short-Lived** because they become clogged with those mineral matters taken up with the soil water which are not used by the plant (64) and which do not pass off in transpiration (75). In most annual plants (337), the older leaves become useless from this clogging and die before the stem is fully developed, and in most perennials the leaves endure but a single season. In the so-called evergreen plants, in which the leaves are usually very thick and are often well protected against evaporation by a very strongly developed cuticle (65), the leaves rarely live more than a few years.

127. The Manurial Value of **Leaves,** that mature on the plant, is usually small, since the more valuable fertilizing materials they contain pass into the stem before the leaves ripen (171). The mineral matters contained in largest quantity by leaves are those that are not used by the plant, but have been deposited within them during transpiration (126).

SECTION IX. THE BUDS

128. The Buds. Each tip of the stem (67) is in most plants protected with a covering of rudimentary leaves or leaf-scales, and the tip with its leafy or scaly covering constitutes a *bud.* A bud forming the apex of a shoot is called a *terminal* bud; one at the junction of a leaf with the stem (axil) is called an *axillary* or *lateral* bud (Fig. 35).

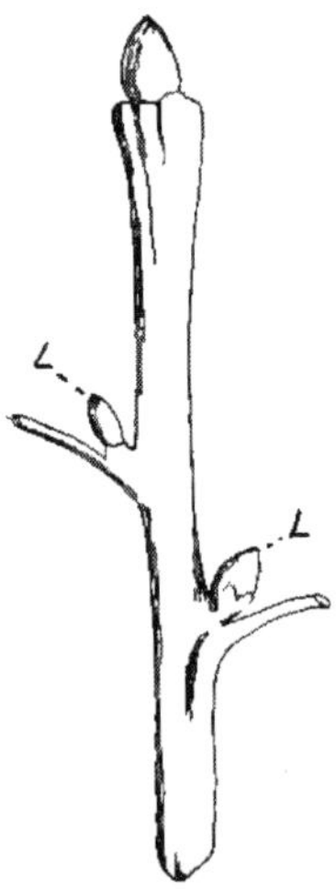

FIG. 37. Buds. L, lateral buds. (After Barry).

Each bud generally includes one terminal- and several axillary growing points. Aside from these, which in the stem exist only in the bud, a bud is simply a part of the stem in which the leaves and internodes are in the embryo stage.

In most perennial plants, the rudimentary leaves that form near the terminus of the young shoots at the latter end of the growing season are changed into *bud-scales,* which serve to protect the growing points within from excessive moisture and sudden changes in temperature. Axillary buds which have not yet formed leaves, are clothed with similar scales. Buds inclosed with scales are often called *winter buds.* To more effectually shut out water, the scales are coated with a waxy or resinous layer in some plants, as the horse-chestnut and balm of Gilead, and to protect them from too sudden changes of temperature, they are lined in other plants, as the apple, with a delicate cottony down.*

129. Nature Provides very **Early** for the **Next Year's Growth** in perennial plants. With the expansion of each leaf, a tiny bud begins to form at its axil, destined if need be to become a branch the following year. Sometimes, however, especially in very vigorous shoots, the embryo buds at the axils of the earliest formed leaves remain undeveloped. The more rapid the growth of the shoot, the less developed, as a rule, are the lateral buds. Cions (386) and cuttings (358) should not, therefore, be taken from excessively vigorous shoots.

130. Branches Develop from **Lateral Leaf-Buds** (132). In trees and shrubs (woody perennials), the lateral buds do not usually push into growth until the spring after their formation, unless the terminal bud is injured. Indeed, they may never push into growth. Some lateral leaf-buds, especially those most distant from the terminal bud, usually remain dormant, through want of light or nutriment, and are overgrown by the enlarging stem the following year. Such overgrown buds, stimulated by destruction or injury of the stem above, sometimes push into growth years after their formation.

---

* A vertical section of the onion bulb may be used as a magnified illustration of a bud as it appears in winter, and that of a head of cabbage, of a bud unfolding in spring.

We can usually decide if detached dormant shoots of trees and shrubs, as cions and cuttings, are of the preceding year's growth or older, since, as a rule, only wood formed the preceding year has visible undeveloped buds.*

131. Adventitious (ad-ven-ti′-tious) **Buds.** Although buds are normally formed only at the nodes of the stem, they may under the stimulus of unusual root pressure (102) be formed without regard to nodes. The trunk of a vigorous elm, willow or horse-chestnut tree, cut off early in the season, often develops a multitude of buds from the thickened cambium (69) at the top of the stump, and a circle of shoots often spring up about the base of a tree of which the top has been injured by over-pruning or severe cold. Such buds are called *adventitious.* It is, however, often difficult or impossible to distinguish between adventitious buds and those that have been previously overgrown (130).

The roots of many plants, as the plum, choke cherry, raspberry etc., develop adventitious buds freely, especially when injured, a fact often utilized in propagation by root cuttings (376).

132. Leaf-Buds and **Flower-Buds.** Buds may contain only rudimentary leaves, or they may contain rudimentary flowers, with or without leaves. The former are called *leaf-* or *wood-buds,* the latter *flower-* or *fruit-buds.* Flower-buds are modified leaf-buds. Both originate in the cambium layer (69) and are normally located at the apex of the stem or in the axil of a leaf (128-129).

---

* Exceptions to this rule are not uncommon in unthrifty trees and shrubs.

133. Flower-Buds are often **Readily Distinguished** from **Leaf-Buds** by location and appearance the same season in which they are formed, which enables the fruit grower to anticipate his crop. In the peach and apricot, and in many varieties of plum, a flower-bud is normally formed on each side of the leaf-bud in the young shoots of bearing trees (Fig. 38). In the apple and pear, the flower-buds are less definitely located, but are mostly formed on the short, thick, wrinkled and crooked branches from wood three or more years old (*fruit spurs*, Figs. 42 and 43). In some fruits, as the apple, cherry and peach, the flower-buds are usually thicker and more rounded

FIG. 38.

FIG. 39.

FIG. 40.

FIG. 41.

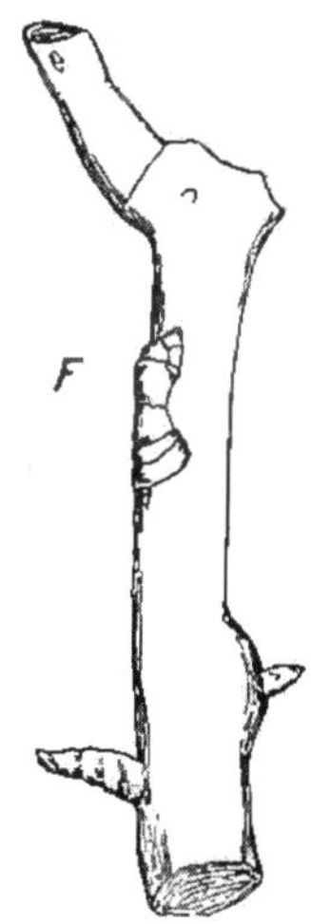

FIG. 42.

FIG. 38. Flower-buds of Pottawattamie plum, *Prunus angustifolia*. The central bud of each group is a leaf-bud.

FIG. 39. Fruiting branch of European plum, *Prunus domestica*. B, young wood. A, wood of preceding year. S, fruit spurs.

FIG. 40. Fruitiug branch of Morello cherry, *Prunus cerasus*. B, young wood. A, wood of preceding year. F, clusters of fruit-buds.

FIG. 41. Leaf-buds of the apple.

FIG. 42. Fruit-bud of apple (F).

All are reduced one-half. (Figs. 39, 40, 41 and 42 are after Barry).

than the leaf-buds, especially toward spring. Close and persistent observation will enable the horticulturist to early distinguish the flower-buds in many of his perennial plants.

In the apple and pear, the buds on the so-called fruit-spurs are not necessarily flower-buds, but some seasons all are leaf-buds. How early in the life of the bud its character is fixed, or if flower-buds ever change to leaf-buds before expanding, does not appear to be known. The fact that leafy shoots sometimes grow out of the center of flowers, and that petals (143) are sometimes developed as leaves, suggest that such a change may occur.

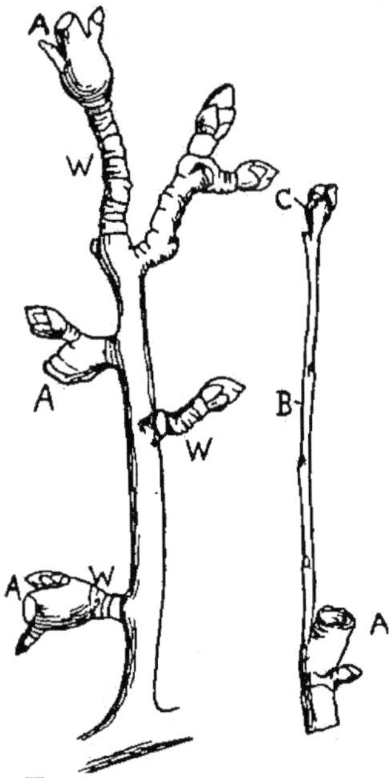

FIG. 43. Fruit spurs of the apple. A, points at which apples were detached the preceding year; W, wrinkles marking points at which fruit and leaves were detached in previous years. Reduced. (After Hardy).

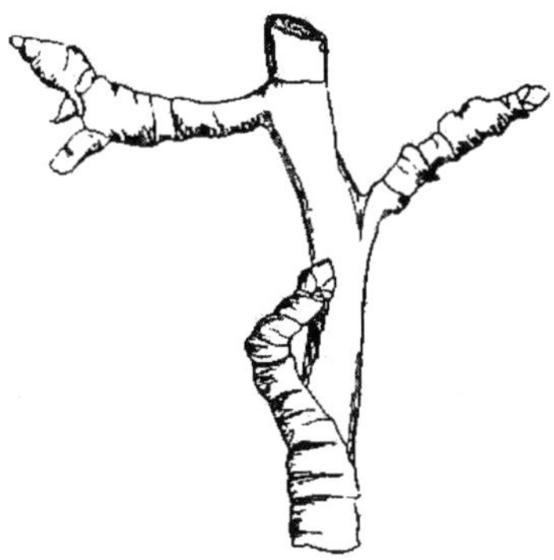

FIG. 44. Fruit spur of the pear. Reduced one-half. (After Barry).

In the grape, flowers appear at the first two, three or four nodes of the young shoots that grow from stems formed the preceding season (*canes*) and the shoot continues to grow beyond the flowers. The raspberry, blackberry and dewberry bloom like the grape, except that the flowers form the end of shoots. In the strawberry, the terminal bud of the preceding year's growth flowers in early spring. In these plants, therefore, the flower-buds are inclosed by the same bud scales that inclose the leaf-buds, hence it is more difficult to foresee the number of flowers than in the tree fruits. A knowledge of the location of the flower-buds is very important in pruning plants grown for their flowers or fruits (416).

134. The Comparative Vigor of **Leaf-Buds** on a given shoot depends somewhat upon their location and the length and diameter of the internodes. The terminal bud, when uninjured, is usually the most vigorous one, and the vigor of the buds, as a rule, diminishes as we recede from the terminal bud. On a given plant, the buds are usually less vigorous on shoots having very long and thick internodes, i. e., the shoots that grew very rapidly (118), than on shoots with internodes of average length and thickness (129). The more vigorous buds are often tenderer than the less vigorous ones, since they are usually farther developed the season in which they are formed.

Cions (386) or cuttings (358), of dormant wood, should be made from shoots having internodes of average length and thickness and with plump, well-matured buds.

In the potato tuber, which is the thickened terminus of an underground stem (Fig. 34), the most vigorous shoot comes from the terminal bud (the so-called *seed-end*), hence rejecting this part of the tuber in planting, as has often been recommended, is detrimental to the crop.

135. Conditions Affecting the **Formation** of **Flower-Buds.** The majority of cultivated plants are grown either for their flowers or the product of their flowers, i. e., fruit or seed. But the flower is not an essential part of the plant, and instead of contributing to its welfare, as do the leaves and roots, it actually consumes a part of the plant's reserve food (140). As might be expected, therefore, perennial plants do not always produce an annual crop of flowers, even when well developed in other directions, hence the grower is often disappointed. Since flowers can only come from flower-buds, a knowledge of the laws that govern the formation of these would often be valuable to the cultivator. Unfortunately, this subject has received less attention than is due to it. Two principles may be cited, however, which if they do not explain all phenomena connected with the formation of flower-buds, are of sufficient general application to have great economic value, viz:

A — *Plants form flower-buds only when they contain reserve food* (85).

B — *A water supply insufficient for rapid growth may suffice for abundant food formation* (59).

In support of the first of these propositions, we mention: (a) Rapidly-growing plants rarely form many flower-buds because the food is used up in growth as fast as formed. (b) Checking such rapid growth, by removing the growing points of the stem or root (67), or by withholding water, results in an accumulation of food and is often followed by an abundant formation of flower-buds. (c) Obstructing the rootward current of prepared food (80), as by "ringing" (416 g) causes an accumulation of food above the obstruction and is often followed by the formation of flower-buds in that part.

In support of the second proposition we mention: (a) Florists often bring their plants into bloom at a desired time by withholding water. (b) The flower-buds of most out-door plants are formed during the drier part of summer,* when a restricted water supply prevents rapid growth, but when abundant sunlight and fully-expanded foliage favor food formation (59).

We may infer, therefore, that *treatment that favors the accumulation of reserve food promotes the formation of flower-buds*, a proposition that is borne out by the experience of practical cultivators.

136. How can we **Promote** the **Accumulation** of **Reserve Food?** Three general principles may be cited:

A — *Provide for abundant food formation* by giving sufficient light and air and by protecting the foliage from attacks of insects and fungi (Chap. III, Section VII).

B — *Provide sufficient plant food* in the soil to satisfy all requirements of food formation (Chap. III, Section VI).

C — *Provide for a moderate check to growth* after the proper amount of growth has been secured.

In the greenhouse where conditions are under control, these principles are readily followed, and the skilled florist rarely fails to secure bloom at the proper time. He gives the desired check to growth by permitting the roots to become densely matted in the pot (*pot-bound*), by withholding water, or by pinching the tips of the more vigorous shoots. With out-door perennial plants, as fruit trees, the problem is more difficult, since conditions are less under control than with plants under glass,

---

* Plants that live over winter and bloom in spring, as the apple, strawberry etc., form their flower-buds the preceding season.

but the principles just cited should always be kept in mind and carried out so far as possible.

We can give sufficient light and air by planting the trees a sufficient distance apart (123) and by proper pruning (Chap. IV, Section III).

If the soil is properly drained, the natural depletion of soil water about midsummer will usually give the needed check to growth. In wet seasons, the drying of the soil may be promoted by stopping cultivation before midsummer and sowing a crop that will increase evaporation from the soil, as oats, clover or buckwheat.

137. Pinching Promotes Flowering (416). In certain cases, as with seedling trees of which we would early know the quality of the fruit, we may give an abnormal check to growth by pinching the tips of the young shoots or by root pruning (416 k). These operations should be performed early in summer, before the period of flower-bud formation, and if the tree is not too young, flowers and fruit may be expected the following season. Frequent transplanting of young trees acts like root pruning, especially if the tap-root is severed. Such harsh measures, however, while they promote early fruiting, doubtless tend to shorten the life of trees.

138. Ringing (416 g) often **Causes** the **Formation** of **Flower-Buds** in otherwise barren trees, by obstructing the rootward current of prepared food. Twisting a small wire about the branch, violently twisting the branch itself, or simple bending and fastening it in an unnatural position, answers the same purpose. But these devices probably weaken the tree and shorten its life by robbing the roots of their normal food supply and are excusable only in special cases, as with seedling trees. It is generally a reproach to the care or knowledge of the cultivator, if his trees of bearing age cannot form flower-buds without such choking.

Fruit trees grafted on slightly uncongenial stocks sometimes flower and fruit more freely for a time than when growing on their own roots, because the imperfect union of cion and stock (383) forms an obstruction to the rootward food-current.

Section X. The Flower

139. The Flower is the developed and expanded flower-bud (132). Its office is to provide for the formation of new plants of its kind (reproduction, 16). Some plants, as the quack grass,* Canada thistle † and horseradish ‡ multiply freely in nature without the aid of flowers, and nearly all plants may be multiplied in culture by other means, but in most of the higher plants, the flower is the natural organ of reproduction, and the only organ devoted solely to this end.

140. Flowers Tend to **Exhaust** the **Plant,** since they are formed from the food prepared by the leaves. But since flower-buds are not usually formed until the needs of growth are provided for (135A), the normal production of flowers does not injure the plant. In certain cases, however, as in plants weakened by recent transplanting or in cuttings (358), flower-buds should be removed as soon as discovered, to prevent their exhaustive influence.

141. The Parts of the **Flower.** The complete flower is composed of four different parts or organs. A knowledge

* *Agropyrum repens.* † *Cnicus arvensis.* ‡ *Nasturtium Armoracia.*

of these parts is of great importance to the botanist in determining species, and also to the plant breeder who would practice cross-pollination (152, 440), hence we need to consider them in detail. The cherry blossom, of which a vertical section appears in Fig. 45, will serve as our first example.

142. The Calyx (ca′-lyx). Beginning at the bottom, the part marked C in the figure, is called the *calyx*. This is green in the normal cherry flower. In some plants, as the flax, the calyx is composed of several distinct, more or less leaf-like parts, each of which is called a *sepal* (se′pal). In the cherry blossom, the sepals are united nearly to the top. The calyx is usually green, but in the tulip and some other flowers it is of another color. In the apple and pear, the calyx becomes a part of the fruit, and its points are visible in the depression opposite the stem.

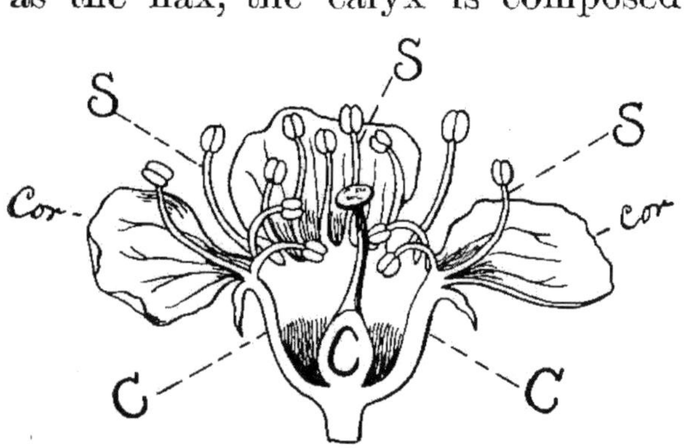

FIG. 45. Section of cherry blossom. C calyx; Cor. corolla; S stamens.

143. The Corolla (co-rol′-la). The more spreading part of the cherry blossom, which is normally white (Cor., Fig. 45) constitutes the *corolla*. In the cherry, the corolla consists of five distinct parts, only three of which appear in the figure, called *petals* (pet′-als). In many plants, as the pumpkin and morning glory, the petals are united. In other plants they are united a part of the way to the top. The corolla is usually of some other color than green.

144. The Stamens (sta′-mens). Inside the corolla is a group of slender organs (S S, Fig. 45) called *stamens.* Each stamen consists of three parts, viz., the long and slender portion, connected with the calyx below, called the *filament* (fil′-a-ment); the swollen part at the top, called the *anther* (an′-ther); and the yellow dust, found upon or within the anther, called the *pollen* (pol′-len). Each grain of pollen is a single cell, which if fertile (153) contains living protoplasm. The pollen is set free at maturity.

145. The Pistil (pis′-til). The column-like part in the center of the flower is called the *pistil.* This also consists of three principal parts, viz., the enlarged flattened summit, called the *stigma* (stig′-ma); the egg-shaped base, called the *ovary* (o′-va-ry); and the slender part connecting the two, the *style.* The ovary contains a smaller, egg-shaped part, called the *ovule* (o′-vule), which when developed becomes the seed. Many flowers have more than one pistil, and many ovaries contain more than one ovule.

Recapitulating, the parts of the flower are, in the order we have considered them:

a — The *calyx;* when divided, the parts are called *sepals.*

b — The *corolla;* when divided, the parts are called *petals.*

c — The *stamens;* the parts are the *filament*, *anther* and *pollen.*

d — The *pistil* or *pistils;* the parts are the *stigma*, *ovary* and *style.*

The ovary contains the *ovule* or *ovules.*

146. The Parts of the **Flower Vary** in **Form** in different species. In the pea flower (Fig. 46) the five petals, shown separately in Fig. 47, are not only quite uulike the petals of the cherry flower, but, as appears, they are unlike each other. The stamens (Fig. 48 st.) and the pistil (Fig. 49) of the pea are also quite different in form from those of the cherry. The variety of form in the parts of the flowers of different species is almost infinite.

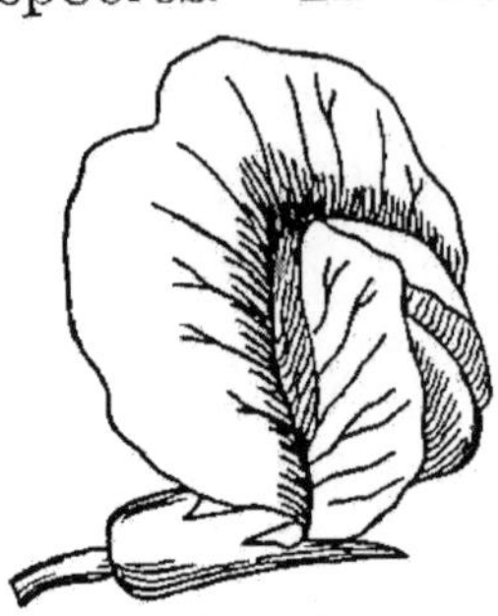

FIG. 46.

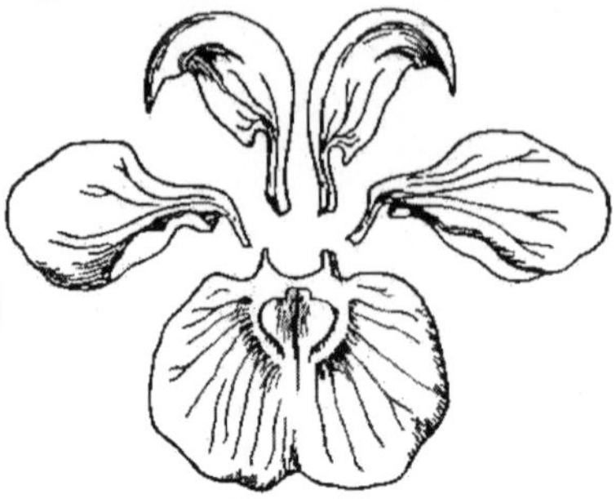

FIG. 47.

FIG. 46. Flower of the pea, *Pisum sativum*. (After Baillon).

FIG. 47. The same dissected, showing variation in form of the petals. (After Figuier).

147. Certain Parts of the **Flower** are often **Wanting.** The flowers of the maple have no corolla; those of the willow have neither calyx nor corolla; certain flowers of the pumpkin, Indian corn and many other plants have no stamens, while other flowers of the same species have no pistils (154). In many varieties of the American plums* the pistil is often wanting.

148. Composite (com-pos′-ite) **Flowers** † are made up of several individual flowers in the same flower-head. The sun-flower (Fig. 50) is a

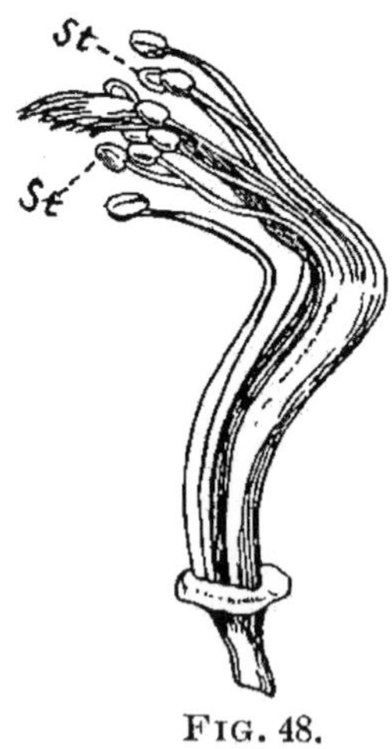

FIG. 48.

FIG. 49.

FIG. 48. Stamens (st) and pistil of the pea, *Pisum sativum*.

FIG. 49. Pistil of same alone. (After Baillon).

* *Prunus Americana, P. angustifolia, P. hortulana.*

† The plants having composite flowers form an extensive family in botany, called *Compositæ*.

familiar example of a composite flower. One of the separate flowers is shown in Fig. 51. At the outer edge of the flower-head, is a row of individual flowers, each of which has a long, yellow, petal-like appendage (Fig. 52), called a *ray.* The flowers bearing rays are called *ray-flowers.* Some composite flowers as the tansy* are without ray-flowers.

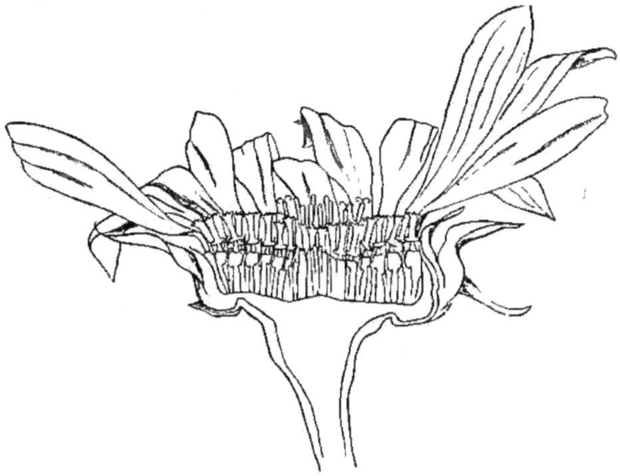

FIG. 50. Cross-section of flower-head of sunflower, *Helianthus annuus.* Reduced. The florets appear closely crowded in the center of the head.

149. The Flowers of the **Grass Family**† to which the cereals belong, as well as corn, sorghum, sugar cane etc., are quite different from those of most other plants. In this family, the flowers are arranged in little groups, each of which is called a *spikelet.* What we call a head of wheat is made up of a number of spikelets, one of which is shown in Fig. 53. Fig. 54 shows the spikelet dissected. The two scale-like parts at the base, g. g., are called *glumes.* The similar pair above, tipped with a bristle (the awn or beard) are called the *lower* or *outer pales* or *palets* (pa'-lets) or *flowering glumes* — to distinguish them from the

FIG. 51.

FIG. 52.

FIG. 51. Enlarged floret of sunflower.

FIG. 52. Ray-flower of same.

* *Tanacetum vulgare.*

† *Gramineæ.*

smaller and more delicate upper or inner palets which are just above and inclosed within the outer palets. Between the outer and inner palets are the stamens and pistils, shown separately in Fig. 55.

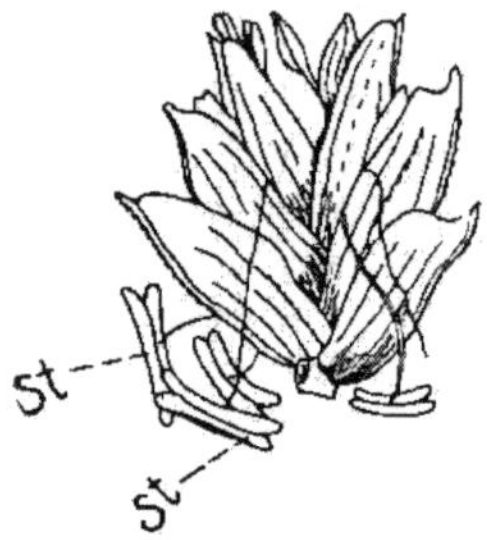

FIG. 53.

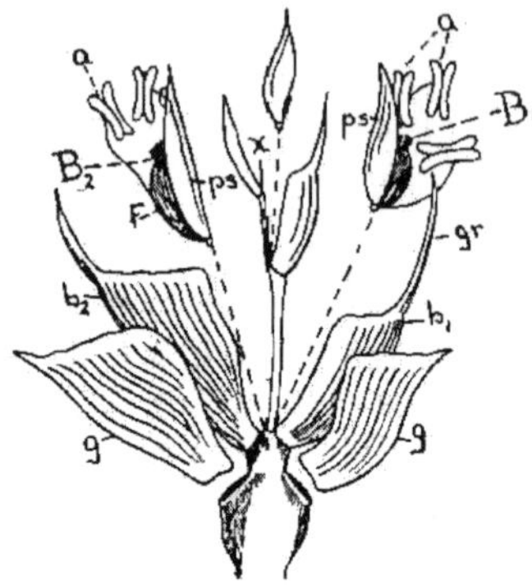

FIG. 54.

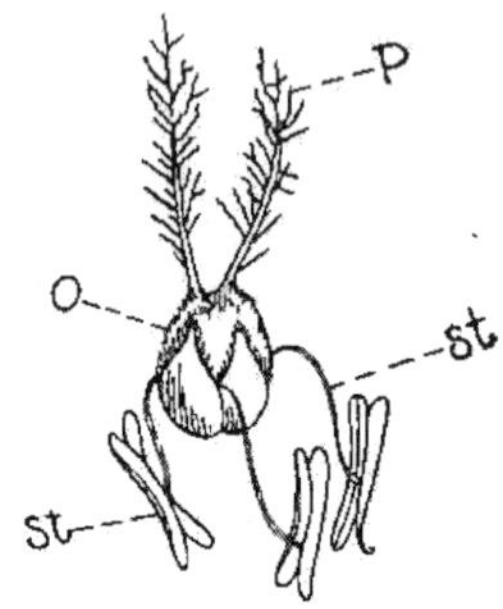

FIG. 55.

Fig. 53. Spikelet of wheat; st, stamens. (After La Maout and Decaisne).

Fig. 54. The same dissected; x, axis of spikelet; g, glumes; b_1, b_2, outer pales; B_1, B_2, flowers displaced from the axis of outer pales; p s, inner pales; a, anthers; f, ovary. (After Prantl).

Fig. 55. Flower of wheat, enlarged; st, stamens; p, pistil; o, ovary. (After La Maout and Decaisne).

150. Fecundation (fec′-un-da′-tion) is the union of the male and female cell by which the new plantlet is formed.* The ovule produces within itself a female cell which may be fecundated by the male cell produced by the pollen (144). This fecundated cell then grows to form a young plant—the embryo (56), and the parts of the ovule develop about it, the whole forming the perfect seed. Unless the ovule is fecundated, the seed very rarely develops. A flower that contains no pistil and hence no ovule, can of course produce no seed.

151. Pollination (pol-lin-a′-tion), is the access of pollen (144) to the stigma (145)—the first step in the process

* The term *fertilization* (fer-til-i-za′-tion), that has been commonly used for this process, tends to confusion, because this term is also applied to the addition of plant food to the soil.

of fecundation. During a certain period, the surface of the stigma is moistioned by the secretion of a viscid liquid, to which the pollen grains readily adhere. Fertile pollen grains,* alighting on the stigma of sufficiently near-related plants during this period, undergo a process comparable to germination, in which a slender projection from the pollen cell penetrates the stigma, passes lengthwise through the center of the style and enters the ovule, where fecundation occurs.

Pollination is not necessarily followed by fecundation. In young plants, and in older plants that are lacking in vigor (9), flowers often fail to produce seed or fruit, even when pistil and stamens appear to be normally developed, and pollination occurs.

In some flowers, as in the pea, the stigma is brought into direct contact with the pollen by the elongation of the style, but in most plants the pollen must be transferred to the stigma by some outside influence, as by insects, the wind, or gravity. Most flowers which have a showy corolla or calyx, or secrete nectar, or yield a fragrant perfume, depend largely upon the visits of pollen-loving or nectar-loving insects for pollination. The showy parts and the perfume serve as signboards to direct the wandering insects to the flowers. Pollination is favored by a warm, dry atmosphere.

152. Cross-Pollination occurs when the stigma receives pollen from a different plant, especially from a plant of a different variety or species (21). The fecundation resulting constitutes a *cross* or *hybrid*, as the case may be

---

* *Fertile* pollen is pollen that is capable of fecundating female cells of its own species.

(23). Cross-pollination is often performed artificially (440).

Close- or *self-pollination* occurs when the stigma receives pollen from its own flower or plant.

153. Cross-Pollination is **Advantageous** in plants, as Darwin's careful experiments have shown. The seeds formed are usually more numerous and larger and make more vigorous plants than with close-pollination. Especially is this true when the parent plants have been subjected to different growth conditions in previous generations. Nature favors cross-pollination in perfect-flowered plants by numerous adaptations tending to prevent self-pollination, as maturing the pollen either before or after the receptive stage of the stigma, or so locating the stamens that the pollen is not readily deposited on the stigma of the same flower.* In some cases, pollen is infertile on a stigma of the same flower or plant that is abundantly fertile on stigmas of other plants of the same species (155).

154. Perfect, Monœcious (mo-nœ'-cious) and **Diœcious** (di-œ'-cious) **Flowers.** Flowers containing both stamens and pistils (or pistil), as in the apple, tomato, cabbage etc., are called *perfect* or *hermaphrodite* (her-maph'-ro-dite); those containing but one of these organs, as in the melon, Indian corn etc., are called *imperfect* or *unisexual* (u'-ni-sex'-u-al).† Flowers of the latter class are called *monœcious* when the stamen-bearing (*staminate* (stam'-i-

---

* Darwin's work "On the Various Contrivances by which Orchids are Fertilized by Insects" describes many most interesting adaptations of this sort.

† The terms hermaphrodite, unisexual and bisexual, though often applied to flowers, are inaccurate.

nate)) and pistil-bearing (*pistillate* (pis'-til-late)) flowers are both produced on the same plant, and *diœcious* when produced on different plants only, as in the hop and date. In a few plants, as the strawberry (155) and asparagus, some individuals produce perfect, and some imperfect flowers.

155. Planting with **Reference** to **Pollination** is important in certain plants. All diœcious plants (154) intended for seed or fruit must have staminate and pistillate plants growing near together or they will not be productive. The hop plant and date palm are of this class.

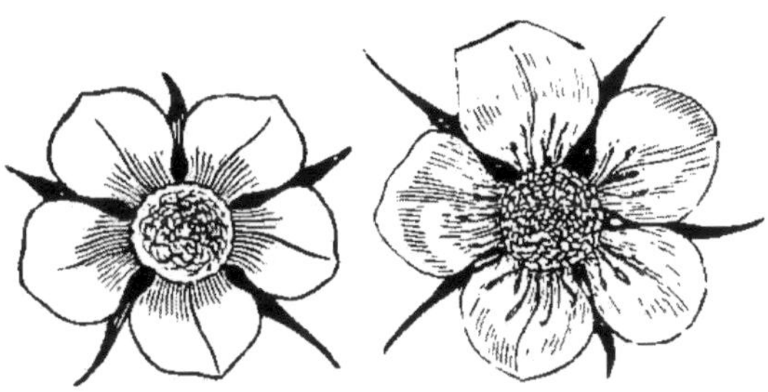

FIG. 56. FIG. 57.

FIG. 56. Imperfect flower of the strawberry.

FIG. 57. Perfect flower of same. The numerous pistils appear in a circular mass at the center, around which the stamens are seen in Fig. 57.

The flowers of many of our most productive varieties of strawberry yield little or no pollen and are unproductive, unless growing near pollen-bearing sorts (Figs. 56, 57). In many varieties of American plums and in certain varieties of the pear, the pollen, even though produced freely, is infertile on stigmas of the same variety. To insure fecundation, *it is wise to mingle varieties in fruit plantations rather than to plant large blocks of a single variety.*

SECTION XI. THE FRUIT AND THE SEED

156. The **Fruit,** as the term is used in botany, is the mature ovary with its contents and adherent parts; it may be hard and dry, as in the wheat and bean, or soft and pulpy, as in the apple and melon. But in common language the term fruit is limited to the pulpy and juicy part of certain plants that normally contains or supports the seed or seeds. To avoid explaining botanical terms, we use the word in the latter sense. In this sense, the fruit serves the plant by attracting animals that can assist in disseminating the seed.

The seed, as we have seen, is the fecundated and mature ovule (145), and its normal office is reproduction (16).

157. The **Fruit Rarely Develops Without Fecundation** of the germ cell of the ovule (150). Varieties of the apple and pear have appeared, however, in which the pulp develops without seeds. The fruit of the banana is almost invariably seedless. The cucumber, grape, orange and fig sometimes develop their fruit without fecundation of the germ cell. These instances are all exceptions to the general rule.

158. Seed Production Exhausts the **Plant** far more than other plant processes. The seed prepares little or no food, while it removes from other parts of the plant a comparatively large amount of prepared food, which it stores up in a concentrated form as a food supply for the embryo (55). Many plants (all annuals and biennials) are killed the first time they are permitted to seed freely, and perennials are often weakened by excessive seeding.*

159. Prevention of **Seeding Prolongs** the **Life** of **Plants.** Many annual flowering plants, as sweet peas, dianthus etc., that soon perish when permitted to mature their

* Double-flowered varieties of the annual larkspur (*Delphinium*), that bear no seed, have become perennial.

seed, continue to bloom throughout the summer if the flowers are persistently picked.* The yield of cucumbers, peas, beans and other garden crops, of which the product is gathered immature, may be considerably increased by preventing the ripening of seed.

160. Overbearing Should be Prevented. Certain varieties of some of our cultivated fruits, as the apple, plum and peach, tend to devote an undue amount of their reserve food to fruit and seed production in favorable seasons, which if permitted, results in enfeeblement or premature death. The wise cultivator guards against this tendency by *thinning the fruit* before it has made much growth, thus saving the tree from undue exhaustion and improving the quality of the fruit allowed to mature.

Thinning should be done as early as the fruits can be properly assorted, and the more imperfect ones should always be removed. The proper amount of thinning will depend upon many conditions, as age and vigor of tree, abundance of crop, fertility of soil, water supply etc. It must be determined by judgment and experience.

161. The Maturing of Seed Injures Fodder Crops. The food value of straw, from which the ripe grain has been threshed, is comparatively small, and that of grass and other crops intended for coarse fodder is much reduced by permitting the seed to ripen before cutting.

162. The Ripening of Fruits. Green fruits assist the leaves in food preparation to some extent, but as they begin to ripen, the process is reversed. Carbonic acid and water are then given off, while oxygen is absorbed.

* See foot note on page 102.

Fruits first become sour from the production of acids which disappear in part at a later stage, while sugar is notably increased. Ripening is favored by warmth and in some fruits by light.

Some fruits, as the strawberry and peach, increase rapidly in size during the ripening period, provided the water supply is sufficient.

Color is not always an index of maturity. Blackberries, currants, and certain other fruits improve in edible quality for some time after assuming their mature color.

Most fruits that have attained nearly normal size, ripen to a degree when detached from the parent plant. Pears are usually improved in quality if picked before maturity and ripened in-doors. The grape, however, fails to develop its sugar if prematurely picked.

After a certain stage of maturity is reached, all vital processes in the pulpy part of the fruit cease, and disorganization (decay) begins, unless prevented by a preservative process.

Section XII. The Gathering and Storing of Seeds

163. The Stage of Maturity at which **Seeds** will **Germinate** varies greatly in different plants and bears no direct relation to the time at which the seeds are set free from the parent plant. Seeds of the tomato will germinate when the fruit is little more than half grown, and those of the pea will germinate when fit for table use. Seeds of the lemon sometimes germinate within the fruit. On the other hand, seeds of the thorn * and juniper rarely germinate until the second spring after their production.

* *Crataegus.*

Seeds of many annual and biennial plants, as the cereals, cabbage etc., may germinate as soon as set free by the parent plant, but those of many annual weeds and of most trees and shrubs will not germinate until some months afterward.

Seeds necessarily gathered immature will often ripen sufficiently for germination if a considerable part of the plant is plucked and cured with them.

Germinating seeds in which the germination process is stopped by undue drying are not always destroyed. Germination may be resumed on access to water. Seeds of different species differ widely in this respect. Those of the parsnip and carrot cannot endure much drying during germination, while those of the cereals may be repeatedly dried at ordinary temperatures without destroying their vitality.

164. Immature versus **Ripe Seeds.** Seeds not fully grown lack a part of their normal food supply, and their embryo is probably imperfectly developed. If capable of germination, they rarely, if ever, produce vigorous plants. As a rule, *the most vigorous plants come from fully-matured seeds.* Immature seeds, persistently used, may tend to reduced vigor, early maturity, dwarfness and shortened life. In some over-vigorous plants, as the tomato, slightly immature seed may tend to increased fruitfulness.

Slightly immature seeds usually germinate sooner than fully matured ones.

165. The Vitality of all **Seeds** is **Limited** by **Age,** but the duration of the vital period varies greatly in different species. Seeds of the chervil rarely germinate if much

7

more than one year old, while those of the gourd family, the tomato and celery often germinate well when ten years old. As a rule, oily seeds, as of Indian corn, rape and sunflower, are shorter lived than starchy seeds, as wheat and rice. The following table gives the average period during which the seeds named are reliable for germination, when properly cared for: *

	Duration of Germinating Power.			Duration of Germinating Power.	
	Average. Years.	Extreme. Years.		Av. Yrs.	Ext'm. Yrs.
Artichoke	6	10	Gumbo or Okra	5	10
Asparagus	5	8	Hop	2	4
Bean	6	10	Kohl-Rabi	5	10
Bean — Kidney	3	8	Leek	3	9
Bean — Soy	2	6	Lentils	4	9
Beet	6	10	Lettuce	5	9
Borecole or Kale	5	10	Maize or Indian Corn	2	7
Broccoli	5	10	Melon — Musk	5	10
Cabbage	5	10	Melon — Water	6	10
Cardoon	7	9	Mustard — Black or Brown	4	9
Carrot	4 or 5	10	Onion	2	7
Cauliflower	5	10	Parsnip	2	4
Celery	8	10	Parsley	3	9
Chervil	2 or 3	6	Pea	3	8
Chervil — Sweet-scented	1	1	Pumpkin	6	10
Chervil — Turnip-rooted	1	1	Rhubarb	3	8
Corn Salad	5	10	Salsafy	2	8
Cress — American	3	5	Sea-kale	1	7
Cress — Common Garden	5	9	Spinach — Prickly-seeded	5	7
Cress — Water	5	9	Squash	6	10
Cucumber — Common	10	10	Strawberry	3	6
Eggplant	6	10	Tomato	4	9
Endive	10	10	Turnip	5	10

166. Conditions Affecting the **Duration** of **Seed Vitality.** A uniform degree of humidity and temperature tends to prolong the vital period of seeds, by causing little drain upon the life of the living cells. Seeds deeply buried in the ground are often capable of germination at a great age, and kidney beans at least one hundred years old, taken from an herbarium, are said to have germinated. In these cases, the seeds were subjected to few variations in humidity and temperature.

* From "The Vegetable Garden," Vilmorin, Andrieux & Cie, Paris.

Seeds usually retain vitality longer when not removed from their natural covering, probably because they are thus exposed to fewer changes of humidity and temperature. Timothy seeds, that become hulled in threshing, lose vitality sooner than those that escape hulling, even when the two sorts have been kept in the same bag. Indian corn is said to retain vitality longer on the cob than shelled, and longer when the ear is unhusked than if husked.

167. Moisture is an **Enemy** to **Stored Seeds,** except for the class that requires stratification (170). A little moisture in stored seeds is very liable to cause the development of fungi (moulds) that may destroy the embryo. Damp seeds are also liable to be destroyed by freezing. *It is important that seeds be dried promptly after gathering,* for if mould once starts, subsequent drying may not destroy the fungus; the latter may resume growth as soon as the seed is planted, thus enfeebling or destroying the embryo before it has time to germinate. Drying by moderate artificial heat (not higher than 100° F.) is wise with seeds gathered in cold or damp weather.

Oily seeds, as of Indian corn, sunflower, and the cabbage family (cabbage, cauliflower, kohl-rabi, Brussels sprouts, ruta-baga, rape, turnip, mustard) cannot safely be stored in bulk in large quantities, except in cool weather.

Seeds are shorter-lived in warm than in cooler climates.

The most satisfactory method of preserving seeds in quantity is to inclose them in bags of rather loose texture and of moderate size, and to store these in a cool, dry and airy place.

168. Age of Seed as **Affecting** the resulting **Crop.** Seeds grown the same or the preceding season produce, as a rule, more vigorous plants than older seeds. They may not, however, in all cases produce plants that are most productive of fruit or seed, for excessive vigor is generally opposed to reproduction. Cucumber and melon plants grown from seed three or four years old are often more fruitful than those from fresh seeds. In crops grown for parts other than fruit or seed, fresh seeds are undoubtedly preferable, but in crops grown for seed or fruit, fresh seed may not always give as large returns as seed of some age. This subject needs further investigation.

169. How Drying Affects the **Vitality** of **Seeds.** The vigor of seeds is probably never increased by drying them, but the seeds of most annual and biennial plants may become air-dry without material loss of vitality. The seeds of many shrubs and trees, however, lose vitality rapidly by such drying and those of some species are destroyed by it. In nature, seeds of the latter class are usually dropped from the parent plant before becoming dry and are soon covered by leaves or other litter that keeps them moist. Nurserymen either plant such seeds as soon as they are ripe, or if of species that do not germinate as soon as ripe, they imitate nature by the process known as

170. Stratification of **Seeds.** This consists in mixing the freshly-gathered seeds with sand, taking care that the sand is kept moist until the time for sowing arrives. Large quantities of seeds may be stratified in boxes, by placing the moist sand and seeds in alternate layers, or the layers may be built up in a pile on the ground. The sand should be coarse enough to admit some passage of air between the particles and to give perfect drainage. The layers should not much exceed an inch in thickness, except for the larger seeds, and the number of layers should not be so large as to prevent proper aeration of the mass. Small quantities of seeds may be mixed with sand or porous loam in flower-pots. Moisture may be maintained in the boxes or pots by burying them a foot or more deep in the soil in a well-drained place, or by storing them in a moist cellar. Care is necessary to keep mice and other vermin from stratified seeds. It is well to cover pots in which valuable seeds are stratified, with a sheet of tin or zinc; metal labels are best for distinguishing different sorts of seed. The seeds should remain stratified until sowing time, when they may be sifted out of the sand or sown with it, as is more convenient. Seeds that do not germinate well until the second spring after maturity (163) are commonly left in stratification until that time.

SECTION XIII. THE DECLINE OF GROWTH AND THE REST PERIOD

171. Annual plants usually perish soon after maturing their seed. In other plants, a certain period of vital activity is followed by one in which growth gradually declines until it almost or entirely ceases. In woody plants, the cells become thickened and a part of the rudimentary leaves change to bud-scales, which inclose the growing point (128). In deciduous (de-cid′-u-ous)* trees and shrubs, the chlorophyll and starch, with most of the

* Deciduous trees and shrubs are those of which the leaves perish at the beginning of the rest period.

potash and phosphoric acid contained in the leaves, are withdrawn into the woody parts (127), while the leaves themselves are detached and fall. The root-hairs also die in many, if not all plants. The leaves of many trees and shrubs assume striking colors as they approach maturity. In perennial herbs, the nutritive matters in the foliage and stem are withdrawn to the underground parts. A period of almost complete repose ensues, during which the plant owing to the dormant condition of its protoplasm is able to endure, without harm, extremes of temperature or dryness that would be fatal in its active state.

Growth ceases in many trees and shrubs earlier than is often supposed. Most orchard and forest trees of mature age grow little, if any, after midsummer in the temperate zones. Cultivation, mulching and manuring tend to prolong the growth period (200).

172. The Rest Period is **Not Peculiar** to the **Temperate Zones,** but occurs in the tropics as well. It can be ascribed in part to the change of seasons, as a few familiar examples will indicate. Tubers of the earlier varieties of the potato, that ripen in northern countries by the beginning of August, do not sprout if left in the ground till October, but if stored in a cellar during winter at a temperature little above freezing, they often begin to sprout in March. Bulbs of the crocus, tulip, narcissus, crown imperial etc., that form in spring, lie dormant in the warm soil during summer and early autumn, but start vigorously in the colder soil of the late autumn or the succeeding spring. The buds of many trees, that form in summer for the next year's foliage and flowers, remain dormant during the hot weather of August and September, to push vigorously in the first warm days of spring. The rest period is to be regarded as a normal, if not a necessary factor of plant life.

173. Most Plants Under Glass Require Rest from time to time, or they do not thrive. The rest is provided either by keeping them at a lower temperature than is favorable to growth, or by submitting them to a degree of dryness that prevents growth. The latter is preferable for plants native in the tropics, where they naturally lie dormant during the dry season.

174. The Time of **Leaf Fall** is an **Index** of **Wood Maturity** in healthy deciduous trees and shrubs. In these, the coloring and fall of the leaves in autumn is not necessarily due to frost, but results from the dormant condition that accompanies maturity. As a rule, the more mature leaves are precipitated by the first autumn frosts. Those less mature usually remain until the more severe frosts. In trees with well-ripened wood, the leaves at the tip of the shoots usually fall before, or not later than, those on the older parts of the tree. With poorly-matured wood the reverse is the case. In a few deciduous trees, as the beech and some oaks, many of the mature leaves remain on through the winter.

175. Hardiness Depends upon the **Degree** to which the **Dormant State** is **Assumed.** Since the most severe climatic extremes come during the natural rest period of plants, the ability of the plant to endure these extremes depends upon the extent to which the protoplasm becomes dormant during the decline of growth. As a rule, a given plant is *hardy* (10) in a locality in which the duration and the warmth of the growing season are sufficient to complete and fully mature its normal amount of growth. Varieties of the apple and other trees, that so far complete their growth in any given locality that their leaves fall before hard frosts, are rarely injured in winter, while those that continue growth until their foliage is destroyed by freezing suffer in severe winters. Deciduous trees are liable to destruction in severe winters in a climate where none of the leaves fall before hard frosts, as is the case with the peach, apricot and nectarine in northern United States.

176. Individual Plants Cannot Adjust Themselves to a **New Environment,** except to a slight extent. The power to complete the annual growth processes and become sufficiently dormant to endure the rigor of the rest period in any given locality is inherited, and not acquired. We are, therefore, able to do very little toward inuring or *acclimatizing* (ac-cli′-ma-tiz-ing) individual plants to an environment to which they were not adapted by nature. We may, however, through the variations of offspring (18), secure varieties in some cases that can endure an environment which the parents could not endure.

177. Plant Processes during the **Rest Period** may not entirely cease. Although food preparation is wholly suspended, root growth and the callusing (73) of injured root surfaces proceed to some extent during the winter in unfrozen layers of soil; and in sufficiently mild weather, the reserve food in the stem gradually moves in the direction of the terminal buds.

178. Cuttings (358) of **Woody Plants** are **Preferably Made** in **Autumn** in climates of severe winters and buried in the ground below the limit of hard freezing, in order that callusing (73) and the transfer of food may make some progress before the final planting.

179. The "Turn of the Year." Toward the close of the dormant season, vegetation, as if benefited by the rest, is prepared to start with renewed vigor, even at moderate temperatures. Buds, that remained dormant during the latter part of the previous summer, push into growth with the first warm days of spring, and many seeds, that could not be induced to germinate the preceding autumn, start with vigor as soon as the soil is sufficiently warm.

The cause for this energetic resumption of plant growth after the rest period is not well understood, but exposure to cold, in the case of temperate plants, and to prolonged dryness in that of tropical ones, doubtless explains it in part, for it is well known that potato tubers may be induced to start their buds soon after maturity by exposing them to the sun a few days, or by placing them for a like time in a refrigerator containing ice. By these means, farmers of Tennessee grow a second crop of potatoes in the latter part of summer and during autumn.

Plants under glass usually thrive better after midwinter than before, and the most favorable time to plant seeds of greenhouse plants is toward the close of the natural rest period.

180. The Round of Plant Life has now been traced, from the first swelling of the planted seed, through the development of the embryo into the plantlet, the penetration of the root into the dark and damp soil cavities, the absorption and conduction of water with its food materials in solution, co-operating with the sunlight and carbonic acid in the mysterious laboratory of the leaf, in building up the plant body into node and internode, leaf, bud, branch, flower, fruit and seed; through growth decline, leaf fall and winter sleep, to the renewed vigor of another springtime.

In our study of the round of plant life, we have assumed the environment to be favorable. But in the practical culture of plants, we are constantly meeting with adverse conditions of environment. Talent for plant culture lies in the ability to discern these adverse conditions by the appearance of the plant, and in knowing how to correct them. We will, therefore, next consider the plant as affected by unfavorable conditions of environment.

The following books are recommended for reading in connection with the preceding chapter: How plants Grow, Gray; Lessons in Botany, Gray; Botanical Text-Book, Gray; A Treatise on the Physiology of Plants, Sorauer; The Soil, King; The Fertility of the Land, Roberts.

CHAPTER III

THE PLANT AS AFFECTED BY UNFAVORABLE ENVIRONMENT

181. Factors of **Environment.** The plant environment is mostly comprehended under the terms, *climate, soil, animals* and *other plants.* But as these are more or less complex influences, it is well to analyze them and to consider separately the component factors of each.

SECTION I. THE PLANT AS AFFECTED BY UNFAVORABLE TEMPERATURE

A — THE PLANT AS AFFECTED BY EXCESSIVE HEAT

182. Transpiration Increases with the **Degree** of **Heat.** The most common effect of heat upon plants is the drooping of the foliage, due to excessive transpiration (75). With insufficient water, this may occur at a temperature that is normal for the plant. But with a water supply that is sufficient at ordinary temperatures, transpiration may be so much increased by an overheated atmosphere that the roots are unable to supply the plant with sufficient water, and as the result, the cells become partially emptied and the foliage droops. Herbaceous plants in an overheated greenhouse or hotbed are sometimes so prostrated from excessive loss of water as to appear dead, but unless the heat has been sufficient to destroy their protoplasm, or the heated period has been protracted, they will recover when normal temperature and water supply are restored.

183. Evergreen Trees are sometimes **Destroyed** by **Untimely Warm Weather** in spring. With a soil so cool that the roots are inactive, a sudden rise of atmospheric temperature, espcially if accompanied by a drying wind, may so far reduce the water in the leaves of evergreen trees as to cause death of the foliage and even of the trees themselves. This most frequently happens in the seed-bed, in compact nursery plantations, or with recently-transplanted evergreen trees. It is most disastrous on poorly-drained clay soils that have a sunny exposure, and at times when the ground is deeply frozen.

The preventives to be observed are, *a*, means that will prevent the tardy thawing of the ground, as thorough drainage and not too close planting; *b*, means that will prevent, in a measure, exposure to the sun, as planting on a northern slope or shading the trees (414); *c*, means that tend to prevent the deep freezing of the soil, as providing wind breaks which tend to retain the snow (204).

184. A Temperature of **122° F.** is **Fatal** to the **Protoplasm** of most land plants. Aquatic plants and the more watery parts of land plants perish at a somewhat lower temperature. Watery fruits, as tomatoes and gooseberries, and the younger leaves of deciduous trees, are sometimes destroyed by full exposure to the sun's rays in very warm weather. An occasional sprinkling of the plants and of the soil about them will usually prevent this result.

185. Plants Under Glass should **Not** be **Sprinkled** in **Bright Sunshine.** Drops of water upon the leaves of plants often act as lenses in converging the rays of the sun, and in a closed greenhouse or hotbed may cause a heat that is fatal to the foliage beneath them. This may explain the brown spots so often observed upon the leaves of indoor plants that have been sprinkled in bright sunlight. Sometimes, but rarely, this trouble occurs in the open air.

FIG. 58. Showing effects of sun-scald on trunk and branches of silver maple tree, *Acer dasycarpum.*

186. Sun-scald is the term applied to an affection of the trunk and larger branches of certain not quite hardy trees, usually upon the south or south-west side, in which the bark and cambium layer (69) are more or less injured (Fig. 58). In severe cases, the cambium is totally destroyed, and the loosened bark splits longitudinally or becomes detached. The effect is apparently the same as when a tree is exposed to the heat of a fire. Sun-scald is most common in young trees and appears to be due, in some cases, to the superheating of the cambium in summer—in others to a return of severe freezing weather after a period sufficiently warm to excite the cambium cells to activity. A preventive is to shade the trunk and larger branches by inclosing them with straw or similar material, or with a lath screen (Fig. 59).

187. Potato Foliage is often **Injured** by **Sun Heat** in summer, as is shown by the browning of the leaves from the tip and edges toward the center, or on the border of holes made by insects. This affection, known as *tip-burn*, is due to the destruction of protoplasm in the cells and is often mistaken for the work of fungus. It is most serious in dry seasons. No remedy for it is known, but it may be in part avoided by selecting varieties least subject to it.

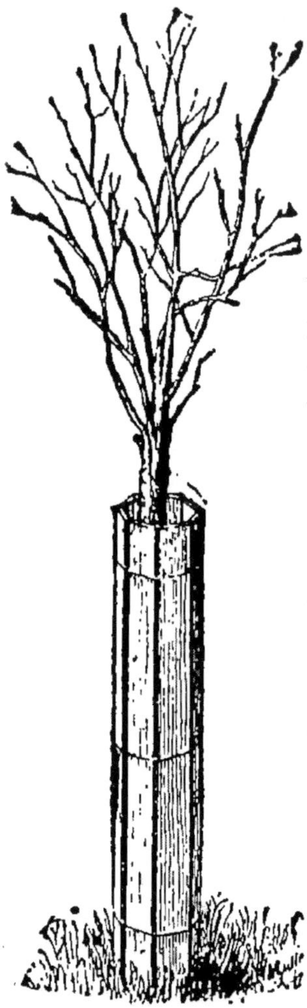

FIG. 59. Trunk of apple tree inclosed in lath screen.

B—THE PLANT AS AFFECTED BY EXCESSIVE COLD

188. The Immediate Effect of **Cooling** the **Plant** is to check the activity of its vital processes. When a certain degree of cold is reached, the protoplasm loses its power to imbibe water (63); hence the plant tissues become less turgid, and the foliage droops somewhat. With a sufficient reduction of temperature, ice crystals form within the tissues and the succulent parts of the plant assume a glassy appearance. The foliage of many plants, as celery, parsnip etc., assumes an abnormal position when frozen.

189. The More Water Plant Tissues Contain, the **Sooner** they **Freeze.** Since the water of plants is not pure, but is a solution of various substances, it does not freeze at the freezing point of pure water (32° F.), but at a lower temperature, determined by the degree of concentration of the solution, or the intimacy with which it is combined with the tissues of the plant. *The more thoroughly dormant the condition of a plant, or part of a plant*, the less water does it contain, and *the better is it able to endure cold* (175).

190. The Power of Plants to Endure Cold depends upon various conditions, aside from the amount of water contained, as

a — *Heredity.* Plants by nature possess widely differing powers to endure cold. The *Anœctochilus* (a-nœc′-to-chi′-lus) perishes when exposed for a considerable time to a temperature of 42° F., while other plants, as the common chickweed,* are uninjured by prolonged. cold, far below the freezing point (176).

b — *The rate of thawing* of the frozen tissues. The more slowly the thawing takes place, the less likely is the frozen part to suffer injury. Many bulbs, tubers and roots which survive the severest winters within the soil, where they thaw slowly, are destroyed by moderate freezing if quickly thawed. Frost-bitten plants are seldom injured when sheltered from the morning sun by a dense fog, which causes them to thaw slowly. Apples, covered in the orchard in autumn by leaves, sometimes pass a severe winter with little harm.

When the water that is withdrawn from the tissues in the freezing process is gradually set free by slow thawing, it may be absorbed by them again and little or no harm results.

c — *The length of time the tissues remain frozen.* A comparatively slight degree of frost, if prolonged, may act

* *Stellaria media.*

more injuriously than a severer degree of shorter duration. Prolonged freezing is especially injurious when the frozen parts are subjected to drying wind, which evaporates their water, while the frozen condition prevents movement of their fluids.

d—*The frequency with which freezing and thawing are repeated.* Frequent slight freezing and thawing are far more injurious than a prolonged frozen condition, even though the latter occurs at a much lower temperature. Winter wheat and rye, and strawberry beds are often more damaged in mild winters, in which freezing and thawing weather alternate, than in more severe ones, when the temperature is mostly below freezing. The chief damage is usually done to these crops in late autumn and early spring.

e—*The previous treatment of the plant.* Plants grown by artificial heat may be far less able to endure cold than others of the same varieties grown in the open air, possibly owing to the more succulent condition of the former. Gardeners *harden* plants grown under glass, by gradually exposing them to the cooler out-door atmosphere, before removing them to the open ground.

f—*The treatment of the frozen tissues.* Handling plants, fruits or vegetables while frozen greatly aggravates the damage from frost, probably because the handling increases laceration of the cells by the ice crystals within them.

191. Frost-Injured Plants, Fruits or **Roots May** often **Be Saved** from serious damage, if promptly placed under conditions that cause the slowest possible thawing of the tissues, as shading from the sun's rays, immersing in ice water or covering with snow. *They should be handled as little and as carefully as possible while frozen.* Sprinkling with cold water is often sufficient to restore frost-bitten plants.

Aside from the death of tender plants by cold, more or less hardy species suffer injury in a variety of ways, of which the following are examples:

192. Destruction of **Terminal Buds** by **Cold.** In plants which do not mature their terminal buds in autumn, as the raspberry, sumac, grape etc., destruction of the tips of growing shoots by frost is a regular occurrence in climates of severe winters. The distance which the shoots are killed back depends upon the succulency of the growth, the severity of the winter and the natural power of the plant to endure cold. Plants thus affected are not always to be regarded as tender, since they often grow wild in climates of very severe winters.

193. The Darkening of the **Wood** (*black-heart*) of certain trees, as the pear, in climates of severe winters, appears to be a chemical effect of the cold. It begins at the center of the stem and in extreme cases may extend clear to the cambium, when the bark ceases to adhere, and the tree or branch thus affected dies. In stone fruits, this trouble is often accompanied by a flow of gum. If the coloring of the wood does not extend to the cambium, the tree or branch may survive, but the first season's growth thereafter is generally feeble and the fruit or the seed crop often fails. During the second season, healthy growth may be resumed, but the heart-wood is rarely or never restored to its normal color. Black-heart often results from other causes than cold, as from bacteria that gain access to the heart-wood through wounds (418).

Other chemical changes result from cold, as the sweetening of potato tubers when chilled, the removal of astringency from the wild grape and persimmon, and the heightening of the flavor of the parsnip.

194. Tree Trunks are sometimes **Split Open** by **Severe Freezing,** the split remaining open until the return of mild weather. This most often occurs in hard-wooded, deep-rooted, deciduous trees, as the oak, and appears to result from the more rapid contraction of the outer layers of the wood in a sudden fall of temperature. The rents are usually overgrown by the next annual wood layer (71).

The splitting down of the main branches of certain varieties of the apple tree appears to be favored by the expansive force of ice in narrow crotches, which retain snow and water. Varieties the branches of which leave the trunk at a wide angle are not subject to this trouble.

195. Bark-Bursting on the trunks of young apple trees often occurs when freezing weather overtakes late-growing and hence poorly-matured wood. In severe cases, the bark splits longitudinally clear through the cambium layer and from the ground to the lower branches; and the bark is loosened from the wood nearly or quite around the trunk. Such trees are practically ruined, but trees slightly injured by bark-bursting may fully recover.

Bark-bursting is usually most severe on deep, rich, moist soil and in seasons that favor late growth, or in which freezing weather occurs unusually early. Late-growing varieties are most subject to it. Its occurrence is lessened by treatment that favors early maturity of the wood (200, 201).

196. Root-Killing of trees. When a very dry autumn passes to winter without rain or snow, the surface layers of the soil sometimes freeze so severely as to destroy the roots of young trees. Root-killing is usually most serious on light soils, and on one-year-old, root-grafted (391) nursery trees, especially when grafted with short cions (386). With very severe freezing on bare ground, root-killing sometimes occurs on soil well supplied with water. The destruction of the roots may be complete or only partial. In the latter case, the tree, if of a vigorous variety, may largely outgrow the trouble, though complete recovery is rare.

Treatment that prevents late growth (200, 201), or mulching the ground about trees tends to avert root-killing.

197. Flower-Buds are often **Destroyed** by **Cold** while other parts of the plant are uninjured. This frequently occurs in the peach, apricot, nectarine and certain species of the plum in climates of rather severe winters, especially after the buds have been somewhat excited by unseasonable warm weather. Flower-buds thus destroyed are dark-colored at the center.

198. Flowers are **Especially Sensitive** to **Cold.** Fruit crops are usually wholly or in part cut off if a slight frost occurs during bloom, and in certain fruits, as the apricot and some species of the plum, the blossoms sometimes appear to be destroyed by a degree of cold that does not descend to the freezing point, possibly through interference with pollination or pollen germination (151). When the freezing is accompanied with snow, however, open flowers may escape without harm, probably owing to the slow extraction of the frost (190 b).

199. Low Plants are often **Destroyed** by **Ice,** especially when the ice layer forms in direct contact with the soil about them and remains for a time after the return of warm weather. The same result is sometimes produced by a covering of snow, of which the top has formed into a crust of ice. Winter grain and strawberry plants are often smothered in this way. Surface drainage of ground devoted to such crops is highly important.

SECTION II. METHODS OF AVERTING INJURY BY COLD

A—DURING THE DORMANT PERIOD

a — *By Treatment of the Soil.*

200. A Dry Soil Favors Wood Maturity, while an abundant water supply retards it. Soil treatment that restricts the water supply toward the close of the growing period tends, therefore, to hasten wood maturity and thus to reduce damage from cold (175). Tillage should be early discontinued about trees liable to winter injury, and in wet seasons, mulching should be removed. Oats, buckwheat or clover sown in the nursery or orchard in late summer promotes wood maturity by increasing evaporation from the soil and is further useful as a covering to the ground in winter (196). Draining heavy or wet soils promotes wood maturity by promptly removing surplus water.

b — *By Treatment of the Plant.*

201. Pinching the **Terminal Buds** (416 a) a few weeks before the time for leaf fall favors wood maturity by checking growth, as does the removal of the younger leaves, in which food preparation is most active. These methods may be employed upon young trees—especially nursery trees, which are very liable to make late growth. Early gathering of the fruit from trees of late varieties also tends to hasten wood maturity.

202. Protection with **Non-Conducting Materials** prevents damage from cold in many herbaceous and shrubby plants in climates where they are not fully hardy. By covering such plants with straw or other litter, or with soil, we lessen to some extent the intensity of the cold, but — more important — we prevent frequent freezing and thawing (190 d), and in a measure, the heaving of the ground, which on heavy or wet soil is destructive to the roots of plants. A covering of straw, leaves or other litter is preferable for low, herbaceous plants, as strawberries. The covering should not exceed an inch or two in thickness, otherwise the plants may be smothered in warm winter weather. For taller plants, as the grape and raspberry, the soil is usually the most convenient and satisfactory covering, as a litter covering tends to attract mice, that often injure woody stems. To assist in bending down the stems, a little earth is usually removed at the base on the side toward which they are to be bent. Shrubs too large for bending down may be inclosed in straw or similar material.

303. A Northerly Exposure is generally **Least Trying** to **Plants** in winter, because it is least subject to fluctuations in temperature. The influence of the sun is here less perceptible and snow remains longer than upon other exposures. The summit of a hill is usually less trying than a valley, because the cold air tends to seek the lower places, especially in still weather (210).

204. Wind-Breaks, i. e., plantings of trees intended to break the force of prevailing winds, act beneficially in lessening damage from cold, in so far as they prevent snow from drifting off the soil and mitigate the effects of drying winds (190 c).

B—Methods of Averting Injury from Cold During the Growing Period

205. Plants are much more susceptible to injury from cold during their growth period than during their dormant period (171). Comparatively few plants, however, are injured by cold at any season until the temperature falls below the freezing point of water (32° F., 0°C.), or when so-called *hoarfrost* occurs. It is this extreme that we have chiefly to fear and to guard against during the growing period.

206. The Cause of Hoarfrost. A sponge saturated with water cannot be compressed in the least unless a portion of the water escapes. If it is but half saturated, it may be compressed somewhat without any escape of the liquid, but if the compression passes a certain limit, the water will begin to escape.

The air is like a sponge in being capable of taking up a certain amount of water. But the amount of water the air can take up depends very much upon its temperature, its capacity for water increasing as the temperature rises, and decreasing as it falls.

Suppose a given amount of air at a temperature of 50° F. has taken up all the water it can hold at that temperature. It is clear from what has just been said, that if the temperature of this air is reduced, some of its water will be set free or *precipitated.* If the air were only half saturated at 50° F., its temperature could be reduced considerably before any of its water would be precipitated; but when a certain degree of cooling is reached, the air will no longer be able to hold all of its water, and a part will be precipitated. The cooling of the air corresponds to the compression of the sponge. The atmosphere always contains more or less water in the form of watery vapor, and the temperature at which any portion of the atmosphere on cooling begins to precipitate a part of its water, is called the *dew point.* The temperatnre of the dew point depends, therefore, upon the amount of water the air contains. When the dew point is above the freezing point of water (32° F., 0°C.), the precipitation is in the form of dew or rain; but when it is below the freezing point of water, the precipitation is in the form of hoarfrost or snow. One more principle needs to be explained, and we are ready to understand

207. How Frost may be **Foretold.** We know that sprinkling the floor of a room cools the air, even though the water used is no cooler than the air of the room. This is because the air in taking up watery vapor absorbs heat, but this heat is set free again when the watery vapor is precipitated. A steam radiator gives out heat because the steam within it is condensing into water. It follows that when the dew point of the atmosphere is reached, a very considerable amount of latent heat is given off, which checks the fall of temperature. *The temperature of still air,* therefore, *rarely falls much below the dew point,* and since the latter at any given temperature depends upon the amount of moisture in the air, if we have an instrument capable of indicating both the temperature and the moisture of the air, we may compute the lowest temperature to which the atmosphere will be likely to descend during any given night, and thus we may foretell frost with some degree of certainty.

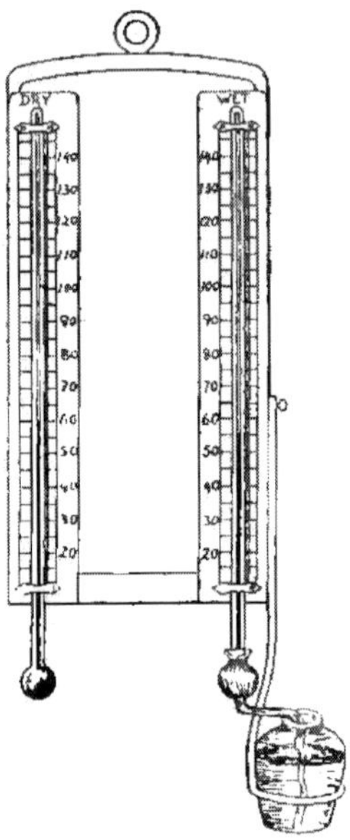

FIG. 60. Sling psychrometer, used to foretell frost.

208. The Sling Psychrometer (psy-chrom′-e-ter) enables us to determine both the temperature and the moisture of the air. This instrument consists of two accurately-graduated thermometers attached to a board or case (Fig. 60). The bulb of one thermometer is inclosed in thin muslin, which is wet just before using the instrument by dipping the bulb in rain water, or is connected with a small vessel containing rain water, as shown. By means of a string attached to the board or case at the end opposite the bulbs, the instrument is whirled about in the air a few times, after which the thermometers are quickly read and the difference in the readings noted. When the air is comparatively dry, evaporation from the muslin proceeds rapidly, and on account of the heat absorbed, the wet bulb indicates a lower temperature than the dry one. When the air is damp, evaporation is slower, and the difference between the readings of the two thermometers is less. In saturated air, evaporation ceases and the two thermometers read alike. By means of the following table, the dew point for any ordinary out-door temperature and atmospheric humidity during the growing season may be readily determined.

Table for Computing the Dew Point in Degrees Fahrenheit.

Dry Bulb.	Wet-Bulb Depression.												
	1	**2**	**3**	**4**	**5**	**6**	**7**	**8**	**9**	**10**	**11**	**12**	**13**
36	32	30	27	24	21	18	13	+8	−1	−12	−32	...	...
37	34	32	29	25	21	19	15	9	+3	− 7	−23	...	...
38	35	33	30	26	23	19	17	11	+6	− 3	−16	...	...
39	36	34	31	28	24	20	16	14	8	0	−11	−31	...
40	37	35	32	29	26	22	18	12	10	+ 3	− 6	−22	...
41	39	36	33	30	27	23	19	14	8	+ 6	− 2	−15	...
42	40	37	34	31	28	25	21	16	10	+ 3	− 2	− 9	−29
43	41	38	35	33	30	26	22	18	13	+ 6	− 3	− 5	−20
44	42	39	37	34	31	27	24	20	15	9	− 1	−12	−13
45	43	40	38	35	32	29	25	21	17	11	+ 4	− 7	−27
46	44	41	39	36	33	30	27	23	19	14	+ 7	− 2	−18
47	45	43	40	37	35	32	28	25	21	16	10	+ 2	−11
48	46	44	41	39	36	33	30	26	22	18	12	+ 5	− 6
49	47	45	42	40	37	34	31	28	24	20	15	+ 8	− 1
50	48	46	43	41	38	36	33	29	26	22	17	11	+ 3
51	49	47	45	42	40	37	34	31	27	23	19	13	6
52	50	48	46	43	41	38	35	32	29	25	21	16	9
53	51	49	47	44	42	40	37	34	30	27	23	18	12
54	52	50	48	46	43	41	38	35	32	28	24	20	15
55	53	51	49	47	45	42	39	36	33	30	26	22	17
56	54	52	50	48	46	43	41	38	35	32	28	24	19
57	55	53	51	49	47	45	42	39	36	33	30	26	22
58	56	54	52	50	48	46	43	41	38	35	31	28	24
59	57	55	53	51	49	47	45	42	39	36	33	29	26
60	58	56	54	52	50	48	46	43	41	38	35	31	28
61	59	57	56	54	52	49	47	45	42	39	36	33	29
62	60	58	57	55	53	51	48	46	43	41	38	35	31

209. To Compute the **Dew Point.** Find the *wet-bulb depression* by subtracting the wet-bulb reading from that of the dry bulb; locate this in the top line of the table, and find the dry-bulb reading in the left hand vertical column. Opposite the dry bulb reading, in the column headed with the number indicating the wet-bulb depression, is the dew point sought.

Example: Dry-bulb reading.........47°
Wet-bulb reading ....40°
Wet-bulb depression..... 7°

Opposite 47°, in the left hand column, and under 7 in the top line, we find 28°—the dew point. If these readings are obtained toward sunset on a clear, still evening, we should expect frost, because the dew point is 4 degrees below the freezing point of water. A slight wind, a hazy atmosphere, or a few fleecy clouds would render frost doubtful. With a dry-bulb reading of 45° and a dew point of 25°, a killing frost might be expected.

210. Cold Air Drainage. Warm air, being lighter than cold air, tends to rise, while the colder air tends to fall. In a still atmosphere, therefore, the colder air accumulates in the lowest places. This explains the familiar fact that hollows and valleys are colder in still weather than ridges and mountains. In a falling temperature and in the absence of wind, gentle currents of the colder air tend to follow the water courses, which explains in part why frost so often "goes in streaks."

211. Wind Tends to Avert Frost because it prevents the settling of the colder air and thus keeps the temperature of the lower strata of the atmosphere nearly uniform.

212. Clouds, Haze and Smoke Tend to Avert Frost because they act to some extent like a blanket in preventing the radiation of heat from the earth, and thus check the fall in temperature (217).

213. The Proximity of a Body of Water Tends to Avert Frost because the water cools slower than the air and thus checks the fall in temperature of the atmosphere in the vicinity; also because it keeps the neighboring atmosphere moist, thereby raising the temperature of its dew point (206). The proximity of buildings and trees tends to avert frost, probably because these objects give up their heat gradually and thus temper the surrounding atmosphere.

214. The Localities Most Subject to **Untimely Frosts** are narrow and deep valleys inclosed on all sides, and inclined valleys that serve as channels through which cold air flows to lower levels. Partially-cleared districts usually suffer more from frosts than those fully cleared, because the remaining forests obstruct air drainage.

Marsh areas are subject to frost, because, in addition to their low situation as compared with the surrounding land, their luxuriant vegetation tends to cool the atmosphere in the vicinity by exposing a large radiating surface and promoting abundant evaporation.

Valleys surrounding elevated lakes that have an outlet through which the colder air may flow to lower regions are particularly free from damaging frosts. The valley of Keuka Lake, in west-central New York, so famous for its vineyards, is of this class.

215. Thermal Belts. In some elevated districts of mountainous regions, localities of greater or less extent are found in which damaging frosts are almost unknown. These have been called *thermal belts*, and their freedom from frost is explained by the merging of the warm air, that rises somewhat rarified by heat from the lower valleys, with the atmosphere of the more elevated region that is rarified to an equal extent by the high altitude. Thus the warm air ceases to rise, but lends its heat to temper the climate of the adjacent mountains.

216. Liability to **Damaging Frost Depends Comparatively Little** upon **Latitude.** Within the tropics are areas where frost is unknown because the temperature never falls to the freezing point. But in localities subject to frost, the liability of damage to vegetation from this cause is governed more by cold air drainage (210) and proximity to water than by latitude. It is as important to select locations for peach growing with reference to spring frosts in the Carolinas as in the peach belt of Michigan, and favorable locations for the apple in Wisconsin sometimes escape damage from spring frosts in seasons when the apple crop is cut off by frost from extensive regions of the southern states.

217. Methods of Preventing Injury by Frost. Any non-conducting material lying between the earth and space, whether spread directly upon the earth or at a considerable height above it, acts as a blanket to intercept the radiating heat, and thus prevents in a measure the cooling of objects beneath it. For this reason, straw, muslin, or other non-conducting material spread over plants, usually protects them from frost.

While it is easy to protect a few plants from frost by covering them directly, it is much more difficult to protect large plantations in this manner. Considerable plantings of the strawberry have been successfully protected from frost by covering the rows in the evening with straw or marsh hay, and where these materials are convenient, the work may often be cheaply and quickly performed.

Attempts to prevent frost on a large scale by the heat of fires or by burning material that produces much smoke or vapor, have not been sufficiently successful to commend these methods for general application.

SECTION III. THE PLANT AS AFFECTED BY UNFAVORABLE WATER SUPPLY

A — BY EXCESSIVE WATER

218. Excessive Water in the **Soil Destroys** the **Roots** of plants. We saw that oxygen is necessary to the life of roots (90). When the soil cavities are filled with water, the roots are soon deprived of oxygen, because the little oxygen contained in the water is soon exhausted (94). Smothering and decay of the roots follow. Seeds planted under such conditions usually fail. The soil water that is most useful to land plants is that which remains attached to the earthy particles after percolation has nearly ceased (capillary water). Such water is well aerated because it is interspersed with cavities that are filled with air.

In the open ground, the remedy for excessive soil water may usually be found in underground drainage. But the same trouble often occurs in potted plants, as the result of too compact soil or too copious watering. The expert recognizes this condition by a sour odor of the soil, by lifting the pot, or by tapping the pot with his knuckle. If the soil is soggy, the weight will betray the fact, or the sound given out by the pot will be that of a compact mass instead of a more or less hollow body, as is the case with a pot of well-aerated soil. To remedy the evil, repot the plant in fresh soil of a proper condition of moisture, providing abundant drainage at the bottom of the pot (412).

219. Injudicious Watering is perhaps the most common cause of failure in growing potted plants. The amateur too often assumes that the chief need of the plants is frequent watering, and so gives water in spoonful doses as the surface soil of the pot appears dry, without observing the state of the soil beneath. The roots of the plants in the meantime may be smothering in water-logged soil or starving from drought. If, owing to inexperience, the condition of the soil cannot be determined by the means above noted, the soil may be tipped out upon the hand without materially disturbing the roots of the plant, by reversing the pot and gently striking its rim on the edge of the bench or table. The real condition can then be readily determined.

220. Copious Waterings at **Considerable Intervals** are **Preferable** to frequent slight waterings. It should never be forgotten that air is as essential as water to the well-being of roots (90), and that the soil, however porous, requires occasional ventilation (94). A considerable quantity of water poured upon the surface soil of a potted plant, in passing downward not only thoroughly moistens the soil particles, but acts like a piston, forcing the vitiated air of the soil cavities ahead of it and out through the drainage hole at the bottom of the pot, while fresh air enters from above as the surplus water passes out beneath.

221. Rapidly-Growing Plants Require More Water and are less liable to suffer from over-watering than slower-growing ones. During the rest period (173), plants should be given very little water.

222. Some Species Require More Water than **Others.** The native habitat of the plant is a partial guide to the amount of water needed. Plants native to arid regions, as the cacti and those from treeless, rocky locations, require little water and are readily destroyed by overwatering. "Plants with narrow and tough leaves, especially when the leaf-blade is vertically placed, do not, as a rule, like much water; plants with broad, leathery leaves prefer a damp atmosphere to great moisture at the roots. Succulent plants with hard epidermal cells (leafless Euphorbias, succulent Composites, Aloes and Agaves), and thin-leaved plants with a strong wooly covering of hairs are further examples of plants which require very little water."*

223. Excessive Watering sometimes **Produces** a **Dropsical Condition** (œdema) in the leaves of plants under glass. This is most likely to occur in winter, when sunlight is deficient, especially if the soil is kept nearly or quite as warm as the air. Water accumulates in the cells, abnormally distending their walls — sometimes even to bursting. An unnatural curling of the leaves, with yellow spots and small wart-like excrescences on their surfaces, are some of the symptoms of this trouble. Less water, increased light and reduced bottom heat (362 a) furnish the remedy.

Frenching, a disease that often attacks growing tobacco on excessively-wet clay soils may be due to the undue absorption of water by the roots. The leaves of affected plants grow narrow, and are thick, fleshy and wrinkled. If the plants are pulled sufficiently to break the tap-root, before the disease has progressed too far, they often recover.

224. Water-Sprouts (sap-sprouts, gormands) on fruit trees are sometimes due to an excess of water in the soil.

---

* Sorauer, Physiology of Plants, p. 207.

These thick, rapidly-growing shoots, with remote leaves and poorly-developed buds, growing from the main branches of unthrifty fruit trees, are most common on undrained, heavy soils. They rarely produce much fruit, but tend to rob the bearing branches of light and nourishment. They usually continue to grow late, and in severe winters are often injured by cold. Water-sprouts may also result from over-pruning and from injury of the tree by cold, but in the absence of these conditions they suggest the need of drainage.

225. Fruits and **Vegetables** often **Crack** from **Excessive Moisture,** either through too much absorption by the roots or by direct absorption through the skin. Cracking is most frequent after heavy rains following drought. Apples, tomatoes, melons, carrots, kohl-rabi, cabbage and the potato tuber are subject to it. On wet soils, drainage may largely remedy the evil. The selection of varieties least subject to cracking is also helpful, especially in melons and tomatoes which often crack in comparatively dry weather. In these cases, the cracking is probably due to an unequal maturing of the fruit which causes certain parts to grow faster than others. The bursting of cabbage heads is due to the excessive absorption of water by the roots. To prevent it, we *start* the plants by pulling on the stem sufficiently to break a part of the roots.

226. Knobby Potatoes are caused by a wet period following a drought during the ripening season. Parts of the plant that are still alive, stimulated by abundant water, resume growth. But since cell division is possible only in the parts containing protoplasm, the mature cells of the tuber can no longer divide, hence growth is limited to the younger parts, i. e., the vicinity of the buds (eyes), and these therefore grow out into unshapely protuberances. The knob consumes a part of the starch previously stored in the tuber from which it grows, hence knobby potatoes are poorer in food value than smooth ones of the same lot.

Certain varieties of potato are more disposed to knobbiness than others. In varieties normally free from it, the planting of knobby seed tubers probably does not tend to increase the inclination to knobbiness.

227. Excessive Moisture in the **Air** is **Injurious** to plants, since it tends to hinder normal transpiration (75) and favors the growth of certain fungous parasites (321). In the greenhouse, we control the atmospheric moisture by ventilation and care in the use of water. Out of doors, we guard against excessive moisture in the air by giving plants sufficient room to favor the circulation of air between them. The latter precaution is especially important in orchard planting, since several fungi that prey upon fruit trees, as the apple scab (328) and pear blight (323), flourish in a damp atmosphere.

B—The Plant as Affected by Insufficient Water

228. Insufficient Moisture in the **Air Causes Excessive Transpiration** (75), which reduces the tension of the cell-walls and thus retards growth (63). It also tends to clog the leaves with useless mineral matters, causing their premature death (126), and favors the development of certain fungous parasites. The effects of insufficient moisture in the air are often very noticeable upon plants kept in living-rooms in winter. Such plants, especially when few in number, rarely make satisfactory growth and the lower leaves continually perish. Moistening the air by evaporating water in the room or setting the pots containing the plants upon a table, covered with moist sand usually remedies the trouble.

Insufficient moisture in the open air rarely occurs unless there is also a dearth of water in the soil.

229. Insufficient Moisture in the **Soil Retards Growth** both by reducing the tension of the cell-walls (63), and by lessening the supply of food from the soil. The tendency of drought is, therefore, to starve the plant.

Plants that have been subjected to insufficient water from the beginning usually suffer less from drought than those previously well watered, because their root system has become more extensively developed (112).

230. Drought tends to **Hasten Maturity,** especially in annual plants, since it favors flowering (135). Lettuce, spinach, rhubarb etc., "run to seed" earlier if insufficiently supplied with water. Potatoes usually ripen earlier in dry seasons than in wet ones. If the drought is sufficiently severe or sufficiently prolonged, diminution or failure of seedage results.

231. Toughness of **Plant Tissues Results** from **Drought.** The crispness and tenderness that give quality to salad vegetables, as celery, lettuce, radish etc., due to a distended condition of their cell-walls, is largely wanting when the water supply during growth has been insufficient.

Insufficient water during growth injures the quality of tobacco. Leaves thus affected have a peculiar spotted appearance when cured, and do not "sweat" properly.

232. Crumbling of the **Surface Soil** (cultivation) tends to **Prevent Drought,** since it greatly lessens the points of contact in the soil particles, and thus interferes with the rise of the soil water by capillary attraction to the surface where evaporation chiefly occurs. An air-dry surface layer of crumbled soil also tends to prevent evaporation by keeping the soil cooler beneath. A puddled crust on the surface of the soil, as is formed by rain on soils containing clay, tends, on the other hand, to restore capillary action and thus to promote evaporation. Some gardeners cultivate their hoed crops as soon as possible after rains for the main purpose of breaking this crust and thus stopping the capillary action.

Cultivation is also beneficial by *aerating* the soil (94). The roots of plants should never be forgotten nor ignored in cultivating crops (110).

233. Mulching tends to **Prevent Drought** by interposing a layer of poor-conducting material between the ground and the sun's rays. This keeps the surface soil cooler and so checks evaporation.

The best mulching material is the one that conducts both heat and moisture slowest. Straw, marsh hay, leaves, manure, shavings, sawdust, spent tan and sand are all useful for mulching, but the first four named are generally preferable to the others, especially if free from weed seeds.

Growing plants tend to dry the soil because the root-hairs continually draw in soil water and force it into the leaves (102) where it passes off by transpiration (75). Weeds, therefore, rob crops of moisture (336).

234. Irrigation, i. e., the extensive watering of outdoor plants, is the final remedy for drought. It is necessary to plant culture in arid regions, and may be profitably employed at certain times in the great majority of seasons in many localities where the annual rainfall would satisfy the needs of crops, were it more uniformly distributed.

235. Drying beyond a certain limit **Kills Plant Tissues** by destroying in part their power for conducting water. Care should be used to retain the normal moisture in buds (394), cuttings (358), and cions (386) and in the roots of plants lifted for transplanting.

SECTION IV. PLANTS AS AFFECTED BY UNFAVORABLE LIGHT

A — BY EXCESSIVE LIGHT

236. The Unobstructed Rays of the **Sun** are often **Injurious** to young seedlings, to unrooted cuttings and to plants recently transplanted. It is difficult to separate the influences of light and heat, since the heat is usually greatest where the sun's rays are brightest; but bright light probably stimulates transpiration (75) independent of heat and thus tends to exhaust the plant of water.

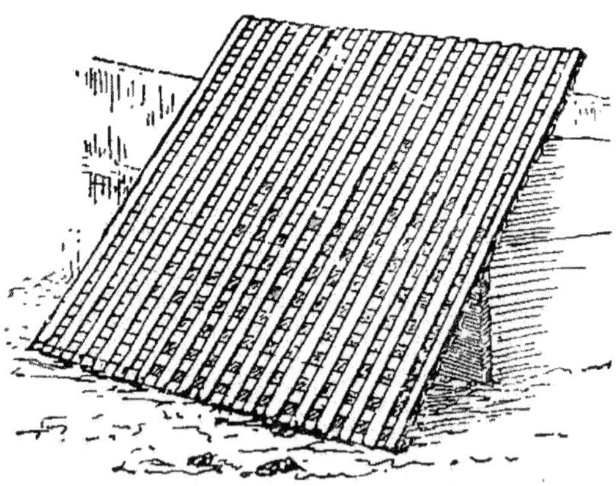

FIG. 61. Lath screen used for shading cold-frames and tender plants in the open ground. (After Bailey).

Various devices are used to break the force of the solar rays. In out-door culture, screens of lath (Figs. 61, 62), cloth or brush (Fig. 63) are often placed over beds containing cuttings or tender seedlings, as of many cone-bearing trees. Cuttings in the nursery may be shaded by supporting a board over the row, on short stakes (Fig. 64), so as to protect them during the warmer hours of the day. Shingles, flower-pots or large green leaves, as of the burdock, are useful for shading plants of the cabbage, tomato etc.

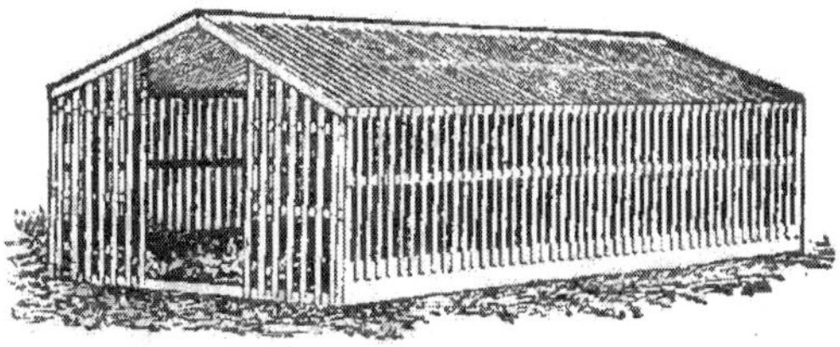

FIG. 62. Shed screen built of three-inch-wide slats, for shading tender plants and for storing pots and boxes of slow-germinating seeds. (After Bailey).

FIG. 63. Brush screen, for shading tender plants in the open ground. (After Bailey).

In culture under glass, the glass is often thinly washed with lime or clay to render it partially opaque, or lath screens are used either above or below the glass. On greenhouse benches, sheets of thin paper or light cloth screens are useful for shading cuttings, recently-planted seedlings and germinating seeds.

FIG. 64. Board shade for recently-set plants, or for cuttings not yet rooted.

Shading should never be so put on as to prevent a free circulation of air about the plants.

A shade that obstructs only a part of the rays of sunlight at a time, as does the lath or brush screen, is generally preferable to one that continuously breaks the force of all the rays, as does paper or whitewashed glass.

237. Cauliflower Heads should be **Sheltered** from **Sunlight** to prevent the formation of chlorophyll in their cells (40), which darkens their color and gives them a strong flavor. The leaves surrounding the head may be tied about it or broken over so as to shade it from direct sunlight.

B—Plants as Affected by Insufficient Light

238. Insufficient Light is a **Frequent Cause** of **Abnormal Development** in plants. Some of its effects are

a—Excessive elongation of the cells of the internodes (76), causing the plants to "draw up" or grow spindling.

b—Deficient formation of chlorophyll (58), giving the foliage a pale-green, yellowish or whitish tint, and resulting in

c—Lessened food formation, causing reduced leaf development and deficient vascular bundles (68).

d—Reduced transpiration tending to watery, weak-celled growth.

e—Weakening of the color and flavor of some fruits, as the apple and strawberry.

f—Preventing pollination (151).

g—Reducing fruitfulness.

Owing to these causes, plants grown in deficient light have tall, slender, weak stems, few, small, pale leaves and scanty roots and are often unfruitful.* Such plants,

* Tomato plants grown in winter on poorly-lighted benches are often unfruitful even when they grow well and bloom freely.

though of species that normally grow upright, are often unable to stand erect without support. Familiar examples are cabbage and tomato plants that lop over when planted out, because grown in the seed-box to transplanting size without "pricking off" (106); and grain sown too thickly on rich ground, that falls (lodges) before maturity.

239. Too Close Planting Causes Deficient Light and all the resulting evils. Indian corn grown too thickly does not ear well and is lacking in nutritive qualities; strawberry plants grown too closely do not fruit well and the fruit lacks flavor and firmness; nursery trees grown too closely are slender-stemmed, deficient in foliage and have poorly-developed roots. A rule to govern distance in planting has already been given (123).

When a slender and flexible growth is desired, as in trees grown for hoop poles, or willows for wicker-work and tying, a certain amount of crowding is advisable.

240. Weeds Cause Deficient Light in low-growing crops as strawberries, dwarf beans, potatoes etc., and also tend to rob the plants of food and moisture. They are, therefore, decidedly injurious (336).

241. Plants Under Glass are **Especially Liable** to **Suffer** from **Deficient Light,** because the walls and sash bars of the structure necessarily intercept a considerable part of the solar rays. The roofs of glass houses should be formed of large lights of glass, with the smallest possible sash bars, and the benches should be arranged to bring the plants as near to the glass as possible.

242. The Electric Light has been found useful as a supplement to the scanty sunlight of short, early-winter days, in forcing certain vegetables and flowers.

243. Insufficient Pruning Prevents the formation of **Fruit-Buds** in orchard trees by restricting light and thus reducing food formation (59). Compare Fig. 65, which shows a fruit branch of the apple tree grown where exposed to abundant sunlight, with Fig. 66, showing one grown in partial shade.*

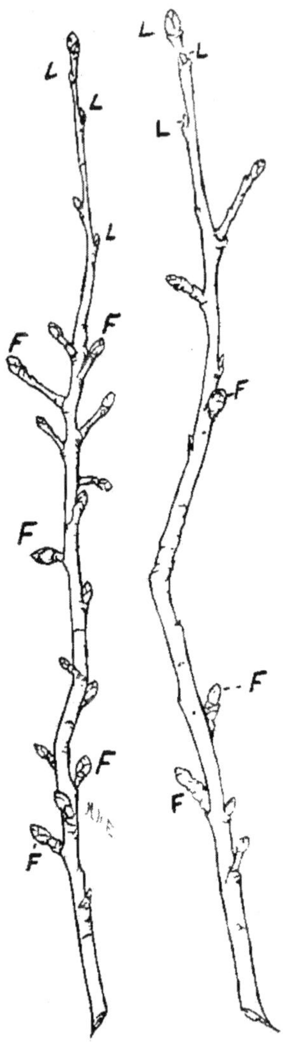

FIG. 65. FIG. 66.

Fig. 65. Fruit branch of apple grown in abundant light.

Fig. 66. Another grown in partial shade.

F, fruit-buds; L, leaf-buds. (After Kinney).

244. Blanching of certain vegetables, as celery, endive, cardoon and sea kale is practiced by gardeners to render them more tender and delicate. It is effected by excluding the light from the parts desired for use, until the chlorophyll (58) mostly disappears, by banking the plants with earth or inclosing them in paper or in drain-tile. Very close planting is sometimes practiced to promote blanching.

SECTION V. PLANTS AS AFFECTED BY UNFAVORABLE WIND

A — BY EXCESSIVE WIND

245. Damage to trees and other plants by excessive wind is too familiar to need notice, except to suggest preventive measures.

a — *The premature blowing off of fruits* may be in a measure prevented by planting fruit trees where they are more or less sheltered from prevailing winds by shade trees, buildings, forests

* See Bulletin No. 37, Rhode Island Agricultural Experiment Station.

or elevations of land. Orchards may be in part protected by planting a wind-break on the windward side (204).

b — *Shade trees in exposed situations* should be headed low, and the head should be formed of numerous branches. The higher the head, the more it is exposed to wind and the greater is the leverage upon the trunk. Several small branches are better able to bear the tempest than a few larger ones.

Shade trees for exposed situations should be of species not likely to be deformed by wind. Certain trees, as the white maple,* often develop one-sided if planted where exposed to prevailing winds, while others, as the sugar maple † and Norway maple ‡ are not thus affected.

B — Plants as Affected by Insufficient Wind

246. Insufficient Wind Promotes the development of certain **Fungous Parasites** (321) by favoring an excessively moist atmosphere. Orchards too closely planted or surrounded by wind barriers suffer more from fungous attacks than those having freer circulation of air between the trees.

247. Insufficient Wind Promotes Damage from **Frost** by permitting cold air to settle in the lower places (210).

On these accounts, gardens and fruit plantations should not be entirely surrounded by wind barriers.

248. Pollination (151) is **Dependent** upon **Wind** in many plants, as the coniferous trees, oaks, elms, birches and sedges; but as the pollen of such plants is very light, their fruitfulness is not often much restricted by insufficient wind.

---

* *Acer dasycarpum.* † *Acer saccharinum.* ‡ *Acer platanoides.*

Section VI. Plants as Affected by Unfavorable Food Supply

We saw that water is the most important constituent of plant food (63) and we have already considered the plant as affected by water supply (Section III). But a proper supply of the other essential food constituents is only second in importance to that of water.

A—Plants as Affected by Excessive Food

249. Excessive food is not the extreme that we have most to fear, since natural soils are rarely excessively fertile, and we can only make them so by costly methods. Indeed, nearly all the constituents of plant food may be present in excess of plants' requirements without working harm. Nitrogen, however, which aside from water is the most potent food constituent, must be used with some discretion.

250. Excessive Nitrogen Stimulates Growth at the expense of flowers, seed and fruit. In crops grown for these parts, therefore, fertilizers rich in nitrogen must be used with caution. Apple, pear and quince orchards liberally manured with such fertilizers produce an excessive, over-succulent growth of wood, that is subject to blight and winter injury and forms comparatively few fruit buds. Grain under similar conditions forms long, weak straw, with poorly-filled heads. Grape vines on over-manured ground produce excessive wood with few and late-ripening bunches.

There is little danger of over-manuring, however, with crops grown for parts other than flowers, fruit or seed, so long as stable manure is used (252). But the more concentrated animal manures, as those from poultry and the hog, the chemical compounds of nitrogen, as nitrate of soda and sulfate of ammonia (262), and the so-called "high-grade" commercial fertilizers must be used with caution, as they may destroy the plants if applied in excess.

B — Plants as Affected by Insufficient Food

251. It is difficult to separate the effects of a lack of food from those of a lack of water, since the food is mainly conveyed to the plant in the soil water (63). But even with a proper water supply, if one or more of the requisite food materials is lacking (61), a normal plant structure cannot be built up. *An excess of one food substance cannot compensate for the lack of another*, except in a few instances.

252. Insufficient Food Dwarfs the **Plant** in all its parts. A dwarfing of the *size* of the plant body may occur, however, without a corresponding dwarfing of the *seed product;* hence plants may often bear their maximum amount of seed or fruit without attaining their maximum dimensions. Plants grown for seed or fruit are, therefore, less likely to be restricted in yield by insufficient food than those grown for their leaves, stems, roots or tubers. The cereals, for example, produce well on land not sufficiently fertile to yield equally good crops of tobacco, cabbage, celery, lettuce or potatoes. But with a sufficient restriction of food, the seed product will suffer diminution or be wholly cut off.

253. Crop-Growing Tends to **Reduce Plant Food** in the soil in proportion as the fertilizing components of the crops are removed from the land and are not returned to it, directly or in equivalent. Fortunately, considerable plant food is constantly being liberated by the disintegration and decay of rock or soil materials, or is being deposited from the atmosphere in rain and snow, so that it is impossible to exhaust the soil of plant food, even with the most improvident culture. But *the cultivator should aim at the largest returns from his soil*, and these are impossible without restoring certain materials that continued crop-removal invariably reduces below the limit of profitable yields.

254. The Food Elements Most Likely to be **Deficient,** when plants are properly supplied with water, are *nitrogen, phosphorus* and *potassium.* These are all liberated in greater or less quantities, when vegetable or animal material (organic matter) decays in the soil; hence all such material has more or less value as fertilizers. But we need not wholly depend upon refuse organic matter for fertilizers, since the leguminous plants add nitrogen to the soil (260), and compounds of nitrogen, phosphorus and potassium may often be purchased in artificial fertilizers at prices that place them within the reach of the cultivator.

255. Nitrogen is the **Most Important Fertilizing Element** because it is liberated in smallest amount by rock decay and is most expensive in the market. Nitrogen is chiefly used by plants in the form of *nitrates,* i. e., in combination with certain other substances as soda, potash, lime, magnesia and iron. Ammonia, which is a gaseous compound of nitrogen and hydrogen, is also used to some extent by plants. Free nitrogen, the most abundant constituent of the air, plays no direct part in plant nutrition (260).

256. The Sources of Nitrates in the Soil are

a — *Nitrification* (nit-ri-fi-ca′-tion), by which the nitrogen contained by organic matter and ammonium sulfate in the soil is changed to nitric acid through the agency of microscopic plants (bacteria). The nitric acid thus formed combines with certain substances (*bases*) in the soil, as potash and lime, forming nitrates (255).

b — *Symbiosis* (sym-bi-o′-sis) on the roots of leguminous crops, through which atmospheric nitrogen is changed to nitric acid (260, 113).

c — *Deposits from the atmosphere* in rain or snow (261).

d — *Ammonium salts* or *nitrates* applied to the soil (262).

257. The Conditions Affecting Nitrification are similar to those affecting plant life in general, since nitrification results from plant life. As it takes place below the surface of the soil, it is favored by the same conditions that favor the root growth of land plants, viz., aeration, warmth and moisture. In general, it is active during the growing season, but at a standstill during the dormant period. It does not proceed rapidly in spring until the soil has become sufficiently warm to promote active root growth.

Nitrification also releases the other food materials contained by organic matter (93).

258. Soil Aeration Promotes Fertility by favoring nitrification. Thus cultivation and drainage (of heavy soils) not only directly promote the growth of plants by assisting aeration (94), but they actually increase plant food. Early plowing in spring promotes nitrification by favoring warming of the soil. Cultivation in dry weather further promotes plant nutrition by preventing the accumulation of soluble plant food in the dry surface soil, where it is deposited above the reach of roots through evaporation.

259. Partially-Decomposed Organic Manures Act More Promptly than fresh ones, because nitrification has already commenced in these materials.

260. Leguminous Plants Enrich the **Soil** with nitric acid (256), which is formed from atmospheric nitrogen in the tubercles on their roots through the agency of microscopic plants (113). Even when a part of these crops is removed from the land, as when clover is harvested for hay or peas for their seed, the land is richer in nitrogen than before the crop was planted. The principal leguminous crops are the clovers, peas, beans, lentils, sanfoin, vetches, alfalfa, lupine and certain species of *Lathyrus.* Highly valuable as are these crops for the nitrogen they leave in the soil, it should be remembered that they do not contribute phosphoric acid or potash, and hence must not be wholly depended upon for soil fertility (263, 264).

Leguminous plants are supplied with nitrogen by the micro-organisms in their roots (113), and hence do not require this element in fertilizers.

261. Rain and **Snow Add Nitrogen** to the **Soil** in small quantities, both as nitric acid and ammonia, which they have taken from the air, but the amounts thus added, while useful to plants, are not under our control.

262. Nitrogen may be **Purchased** for fertilizing purposes as *sodium nitrate* (nitrate of soda, Chili-saltpeter), *ammonium sulfate* (sulfate of ammonia), and in organic materials. The former is available as plant food when it is dissolved in the soil water. It is best applied immediately before the planting of a crop or in small quantities at intervals during growth, since it is in danger of being washed out of the soil in drainage water. Sodium nitrate is especially useful for garden crops started early in spring, when the soil is too cool for active nitrification (256). The surface soil is apt to be poor in nitrates in spring, because they are often washed down by the autumn and winter rains.

Ammonium sulfate is changed to nitrates in the soil before it is used by plants, and hence is less prompt in its action than sodium nitrate. It is more tenaciously held by the soil than sodium nitrate and is therefore less likely to be lost by washing.

263. Phosphorus is used by plants in the form of soluble phosphoric acid, which exists in the soil in combination with lime, iron and alumina, as phosphates of these substances. It may be purchased in the form of mineral phosphate of lime, ground bone, wood ashes, odorless phosphates etc. The first two are insoluble in water unless treated with a strong acid, when they are known as *acid phosphate* or *superphosphate.* Phosphoric acid is not readily washed out of the soil, even in its soluble form.

264. Potassium is used by plants in the form of potash, i. e., potassium combined with oxygen. Potash exists in the soil mainly in combination with chlorin (chlorid or muriate of potash), with sulfuric acid (sulfate of potash), or with nitric acid (nitrate of potash). All these forms of potash are freely soluble in water and are immediately available as plant food. Nitrate of potash (saltpeter) is a most valuable fertilizing material, since it contains both potash and nitrogen, but unfortunately its price is too high to permit its use for this purpose. The muriate, either pure or in crude form (kainit), and sulfate may, on the other hand, be purchased at reasonable prices. The sulfate is considered preferable for tobacco and potatoes as it is thought to produce a better quality of product. The muriate acts more promptly than the sulfate, however.

265. Wood Ashes are a **Valuable Fertilizer,** especially when unleached, as they contain both potash and phosphoric acid. In leaching, the potash is mostly washed out, but the phosphoric acid is largely retained. Ashes contain no nitrogen.

266. Farm and **Stable Manures** should be the first dependence of the cultivator. Aside from these, leguminous crops (260) are undoubtedly the cheapest source of nitrogen for the farm, and with unleached wood ashes, furnish all the needed fertilizing ingredients for grain crops grown in rotation. For garden crops, however, if sufficient stable manure can not be obtained, more nitrogen may often be profitably used than can be furnished by leguminous crops, hence for these, commercial fertilizers may often be added with advantage.

267. Crops Suggest Their Own Needs to some extent, so long as they are not suffering from drought. As a rule, a lack of nitrogen is indicated by pale-green foliage or small growth of leaf or stalk. Excess of nitrogen is indicated by very large growth of leaf or stalk, with imperfect bud-, flower- and fruit development. Lack of phosphoric acid is indicated by scanty crops of light or shrunken seed on plants of normal size. Lack of potash is indicated by small crops of inferior fruit, accompanied by satisfactory growth.

268. Crop Rotation Economizes Plant Food, because some crops use more of a given food constituent than others. The alternating of crops having different food requirements tends to prevent the exhaustion of special food substances.

269. A Growing Crop Tends to **Conserve Fertility,** because it reduces drainage by taking up water from the soil, and at the same time, appropriates the available plant food, thus preventing loss of the latter from leaching.

270. Manures are, in part, the **Raw Material** from which the cultivator turns out valuable products. They should, therefore, be most carefully preserved and applied. Leaching of the manure pile by undue exposure to rain and over-rapid fermentation, by which nitrogen escapes as ammonia or other gaseous nitrogen compounds, should be stringently avoided. All refuse organic matter should, so far as possible, be made to increase the always-too-small stock of manure.

Section VII. Plants as Affected by Parasites

The only instance of a beneficial plant parasite (24) of special interest to the cultivator, is the micro-organisms in the roots of leguminous plants, which we have already considered (260). Many parasites of harmful insects are beneficial, but these are beyond our scope. We need therefore to treat here only those parasites that are directly injurious to economic plants.

10

271. The Injurious Parasites of plants are **Very Numerous** and a scientific classification of them is beyond the limits of this work. We shall only endeavor to arrange the different parasites into groups based on their manner of working injury and the methods by which they may be controlled.

With reference to the character of their injury and the preventives used, as well as in their natural characteristics, plant parasites are readily separable into two great classes, viz., animal and vegetable parasites. These classes will be considered in their order.

A — Plants as Affected by Animal Parasites

a — By quadrupeds and birds. The four-footed animals that injure cultivated crops nearly all belong to the class known as rodents, which includes mice, gophers, rabbits, woodchucks, moles etc. These may usually be controlled by trapping, shooting, or poisoning, or by protecting the plants.

272. Damage from **Mice** to orchard and nursery trees is very common. Mice are usually most troublesome on sod ground covered with snow, especially beneath snow banks, hence all grass should be removed in autumn from the immediate vicinity of the trees. It is well to ridge the soil a little, directly about the trees, so that the mice in burrowing beneath the snow will not be likely to come in contact with the stems. Packing the snow immediately about the trees is helpful when damage is discovered during winter. The stems of orchard trees may also be wrapped in heavy paper or inclosed in fine wire netting. If tarred paper is used, it should be promptly removed in spring, or it may injure the bark.

Stored seeds of almost all kinds must be carefully guarded against mice.

273. Gophers are often troublesome by eating planted seeds and by burrowing about the roots of young orchard trees. They may be poisoned by placing corn, soaked in a weak solution of strychnin in water, about their holes.

274. Rabbits are especially troublesome to nursery trees, when the ground is covered with snow. The most satisfactory protection is to inclose the nursery with a fence of poultry netting, which should be banked up a little at the bottom to prevent the rabbits from passing under. It should be high enough to reach above the surface of the deepest snow.

Orchard trees may be protected against rabbits by inclosing the trunks with the devices mentioned under sun-scald (186). Smearing the stems with blood has also been recommended.

275. Woodchucks are often troublesome to growing crops, but as they are seldom numerous, shooting or trapping generally suffices to prevent serious damage. Moles are very troublesome in some localities by eating the roots of plants. They may be largely controlled by the use of mole-traps. Pouring a little carbon bisulfid into their holes is also generally effectual.

276. Birds are often troublesome by eating unharvested fruits and seeds. Inclosing the trees or plants with fish netting, when this is practicable, is perhaps the most satisfactory preventive. The netting is not expensive, and the same piece may be used several seasons.

b—By insects, worms, slugs and snails. As worms, slugs and snails work the same kind of injuries as some insects and are controllable by the same methods, we do not distinguish between them in the following paragraphs.

277. Many Insects are **Beneficial** by destroying harmful insects or by promoting pollination (151). We should not, therefore, wage indiscriminate warfare upon all insects.

278. Methods of **Preventing Insect Ravages** to plants are various, as inclosing the plants, trapping, repelling or removing the insects, destroying them by means of insecticides, or preventing reproduction by destroying the eggs. The important question in the case of any injurious insect is by which one of these methods it may be most effectually and cheaply controlled.

279. Inclosing the **Plants** is practicable in a few cases, as with the striped cucumber beetle.* The hills in which cucumbers, melons, squashes etc., are planted, may be covered with a frame having fine-meshed wire- or cotton netting tacked over the top, which prevents the beetles from gaining access to the plants (Fig. 67).

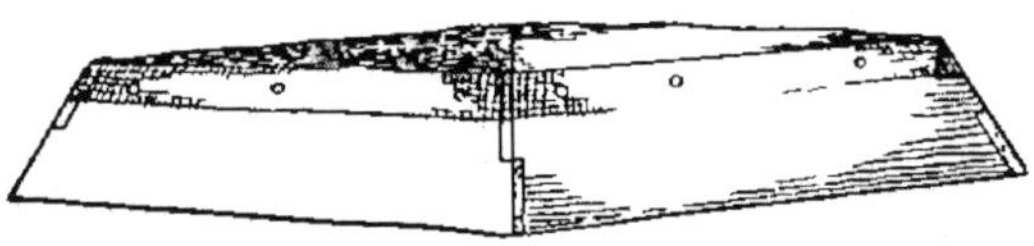

FIG. 67. Screen-covered frame for protecting hills of the melon and cucumber.

280. Trapping the **Insects** is practicable in a few cases, as with cutworms, which often conceal themselves during the day beneath objects on the ground. They will frequently be found in numbers beneath handfuls of green clover or other herbage placed on the ground near the plants which it is desired to protect. By poisoning the herbage (284), some of the cutworms may be killed, but many are likely to escape unless destroyed by other means.

* *Diabrotica vittata.*

281. Repelling Insects by means of offensive odors is partially effectual in some cases, as with the squash-vine borer.* Corncobs or other objects, dipped in coal tar and placed among the plants, repel many of the moths that produce the borers.

282. Hand Picking, i. e., removing the insects from the plants by hand, is the most satisfactory method for destroying certain insects, as the tobacco- or tomato worm,† and other large caterpillars, and the rose-beetle.‡ A vessel of water with a little kerosene on the surface, in which to throw the insects as they are gathered, is a convenient way of destroying them. In some cases, the insects can be shaken or knocked from the plant directly into the vessel. This method is often employed in destroying the potato beetle.§ Digging out cutworms and white grubs‖ from about corn and strawberry plants, and cutting out borers from trees and squash vines are often the most effectual methods for destroying these insects.

283. Destroying Insects by **Poisons** or **Caustics** is the method most generally available. The material used is called an *insecticide* (in-sect′-i-cide), and if satisfactory, must be destructive to the insects without injuring the plant to which it is applied, or rendering the plant or its products unfit for food. The insecticides in most general use are certain compounds of arsenic (Paris green, London purple, white arsenic), hellebore and pyrethrum powders, tobacco, kerosene and various compounds of soda and potash. With the exception of kerosene and

* *Melittia ceto.* † *Phlegethontius celeus.* ‡ *Macrodactylus subspinosus.*
§ *Doryphora decemlineata.* ‖ *Lachnosterna.*

the soda and potash compounds, all these may be used either as a dry powder or with water.

284. The Arsenic Compounds are effectual as insect destroyers, even when largely diluted. When applied in water, however, they are liable to injure foliage in proportion to the amount of soluble arsenic they contain. When insoluble in water, they require stirring to keep them in suspension.

285. Paris Green (arsenite of copper), when pure, is a nearly insoluble compound and may be safely used upon the foliage of most plants, diluted at the rate of one pound to two hundred gallons of water. For the peach and nectarine it should be diluted one-half more. Pure Paris green dissolves without sediment in ammonia water.

286. White Arsenic (arsenious acid) is slightly soluble in water, and hence is dangerous to foliage unless used with care. If applied immediately after its addition to the water, it may be safely used upon the foliage of the apple, plum and cherry at the rate of one pound to fifty gallons, but constant stirring is required to keep it in suspension.

287. London Purple (arsenite of lime, with certain impurities) often contains soluble arsenic, and like white arsenic, must be used with caution. It may be safely applied to many plants at the rate of one pound to two hundred gallons of water, if put on immediately after its addition to the liquid, but for the peach it should receive greater dilution. London purple is considerably cheaper than Paris green.

The addition of fresh milk of lime to water, to which white arsenic or London purple has been added, largely prevents their tendeucy to injure foliage.

Both Paris green and London purple, when perfectly mixed with 150 parts, by weight, of land plaster, or an equal bulk of any other cheap, harmless powder, are satisfactory for destroying the potato beetle and many other leaf-eating insects (307).

288. Compounds of **Arsenic** are **Deadly Poisons** and should always be handled with the greatest care.

289. Hellebore (hel′-le-bore) **Powder,** i. e., the ground root of white hellebore,* is a far less virulent poison than the arsenic compounds. It is therefore useful in destroying a class of insects against which a deadly poison cannot wisely be used, as the imported currant worm,† and the cabbage caterpillar.‡

Hellebore powder when used dry, may be diluted with once or twice its bulk of flour, which causes it to adhere better to the foliage than if used alone. When applied with water, a heaping teaspoonful or more may be added to three gallons. The dry powder is very light and should only be used in a still atmosphere.

A decoction made by boiling the root of white hellebore in water is said to possess insecticide properties similar to those of the powder.

290. Pyrethrum (py-re′-thrum) **Powder,** (Persian insect powder, Dalmatian insect powder, Buhach) is the pulverized flowers of certain species of Pyrethrum.§

Pyrethrum powder is not poisonous to the higher animals, but the oil that pervades it is destructive to many

---

* *Veratrum album.* † *Nematus ribesii.* ‡ *Pieris rapæ.*

§ "Persian insect powder" is made from the flowers of *Pyrethrum roseum* and *P. carneum*; "Dalmatian insect powder" and "Buhach" are made from those of *P. cineraiæfolium.* "Buhach" is the trade name of a pure product prepared in California.

insects. As the oil is extremely volatile, pyrethrum is better adapted for use under glass or with plants otherwise inclosed. It is not injurious to foliage or flowers. Fresh and pure pyrethrum powder may be diluted half or more in bulk with any other light, cheap, harmless powder, but the mixture should stand a day or two before use, to enable the diluent to absorb the oil. The powder may be used with water at the rate of half a pound to three gallons.

The pyrethrum plant is comparatively hardy and has been successfully grown in northern United States. It is said that a decoction of the unopened flowers possesses the insecticide properties of the commercial product.

291. Hellebore and **Pyrethrum Powders** should be **Kept** in **Close Vessels,** since their poisonous properties are volatile. In purchasing, only fresh samples should be accepted. If fresh and pure, these powders produce a tingling sensation when applied to the nostrils.

292. Tobacco Smoke is much used for destroying "lice" or "green fly" (aphidæ) on plants under glass. For this purpose, the partially-dry stems or leaves are burned upon pans or bricks, or in small sheet-iron stoves. Many delicate flowers are, however, injured by tobacco smoke.

Stems or leaves of tobacco, strewn abundantly beneath greenhouse benches, tend to prevent the multiplication of aphidæ.

Several semifluid extracts of tobacco are sold which may be evaporated in the greenhouse over an oil stove, or preferably by steam under pressure. Some of these are very efficient for destroying insects and do not injure flowers.

293. A Strong Decoction of **Tobacco** is often used for destroying aphidæ on plants in rooms where tobacco smoke would be objectionable. The plants are immersed in, or washed with the decoction. The same is often effectually used on young plants of cabbage, cauliflower and turnip, to prevent their destruction by the flea beetle.*

294. Kerosene is a very useful insecticide for a class of insects not readily destroyed by other means (316). It has generally been used as an emulsion made with soap and water, for which the following formulas are good.

a — Dissolve one quart of soft soap, or one-fourth pound of good hard soap, in two quarts of boiling water; remove from the fire, and pour into a tin can containing one pint of kerosene. Agitate violently in the can for three minutes. For use, dilute with an equal quantity of water; or

b — Dissolve one-half pound of hard soap in one gallon of boiling soft water; add at once two gallons of kerosene, and churn or otherwise violently agitate for five or ten minutes. For use, dilute with 15 parts of soft water.

Kerosene may also be applied in intimate mixture with water, secured by pumping both liquids at once through a good spraying nozzle (Fig. 70). About ten per cent of kerosene should be used for most plants.

295. Caustic Potash in solution is useful for destroying certain scale insects, as the oyster-shell bark-louse,† for which solutions of one-fourth pound to the gallon of water may be applied during winter.

---

* *Phyllotreta vittata.* † *Mytilaspis pomorum.*

296. Resin (or rosin) **Washes** are valued for destroying various scale insects in southern and western United States. They are adapted with modifications to both dormant and growing trees. The resin is sometimes saponified with caustic soda and simply diluted with water; fish oil or petroleum may also be added. The following and other formulas are in use:

a — Dissolve one pound of caustic soda in one gallon of water in a covered iron kettle. Pour out half of the solution, and to the remainder add 8 pounds of resin and boil until dissolved. Then pour in very slowly the rest of the caustic soda solution and boil the whole, stirring it constantly, until it will unite with water, forming a liquid resembling milk. Dilute to 22 gallons for use. This mixture may be applied during the growing season; or

b — Place 30 pounds of resin, 9 pounds of 70 per cent caustic soda and $4\frac{1}{2}$ pints of fish oil, in a closed iron kettle and cover with five or six inches of water. Boil until the liquid has a dark-brown color, after which slowly add water until the whole makes 100 gallons; or dilute a part of the liquid at this rate, keeping the remainder as a stock solution. This is for use in the dormant season. For the growing season, similar, but more dilute solutions are used.

297. Hydrocyanic Gas. Another method of destroying scale insects used in California and the Southern States is to treat the tree, previously inclosed in an oil-cloth tent, with hydrocyanic gas. One ounce of cyanid of potassium and one measured ounce of sulfuric acid are placed in an earthern or leaden jar containing three fluid ounces of water. The jar is covered with burlaps to prevent the rapid escape of the gas. The tent is left over the tree fifteen minutes to one hour. It is advisable to apply this treatment during the dormant season and in a cool period.

298. Fir-Tree Oil is considerably used in greenhouses and conservatories for destroying scale insects and the mealy bug.* It is mixed with warm soft water at the rate of a tablespoonful of oil to a pint, and applied with a syringe; or the plauts are dipped into the mixture.

299. Hot Water may also be used for destroying the above-named insects (298) and plant lice (aphidæ). Infested pot-plants are inverted and immersed five or six seconds in a vessel containing water at 120° F. This treatment must be used with caution.

300. Insect Attacks Sometimes Become Formidable from the vast number of the individuals. The chinch-bug,† the army-worm ‡ and various species of locusts or grasshoppers sometimes devastate large tracts of country. For the destruction of these insects, special means must be employed.

301. The Chinch-Bug may be controlled in a measure, by burning over all grass land in early spring, in seasons when attacks are expected. The bugs may be kept out of corn fields by plowing a furrow away from the corn, on the side from which the attack is looked for, and strewing stalks of fresh corn in this. As the insects congregate on the corn in the furrow, they should be destroyed with kerosene (294). Persistent and thorough work is essential to success.

---

* *Dactylopius.* † *Blissus leucopterus.* ‡ *Leucania unipuncta.*

302. The Army-Worm may often be prevented from migration by plowing a deep furrow, as above directed, and making the side toward the endangered crop vertical, with a spade or shovel. The insects will congregate in the furrow where they may be destroyed by dragging a log over them.

303. Grasshoppers and **Locusts** may be destroyed, before they have attained their wings, by drawing over the infested ground, a "hopper-doser" which consists of a shallow, sheet-iron pan, with a vertical, cloth-covered back. The pan contains a little kerosene, and the cloth back is kept saturated with the same liquid. The insects jump into the pan or against the cloth back, thus becoming wet with the kerosene, and soon perish. Grasshoppers may also be poisoned by distributing bran mixed into a mash with water containing arsenic in solution. Plowing grass land containing the eggs of grasshoppers tends to prevent an attack.

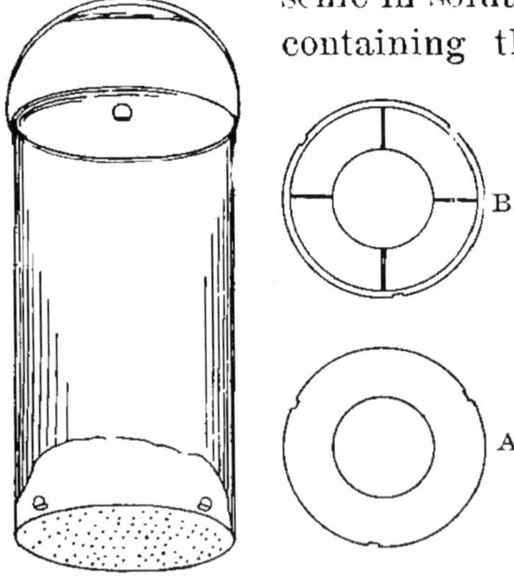

FIG. 68. Sifting box for applying powders.

304. Apparatus for Applying insecticides. *Powders* are readily applied upon low-growing plants, as the potato, cabbage etc., by means of a sifting box consisting of a pail with a perforated bottom, a rigid handle and a tight-fitting cover (Fig. 68). For small plants, as young potato tops, the tin disc A which has a circular hole in the center, is laid inside on the bottom of the box, and held in place by small lugs soldered to the wall as shown. When it is desired to spread the powder more, the disc B is used.

For taller plants, a powder bellows is desirable.

Liquids are best distributed with a force pump, fitted with a hose of a length suitable to the height of

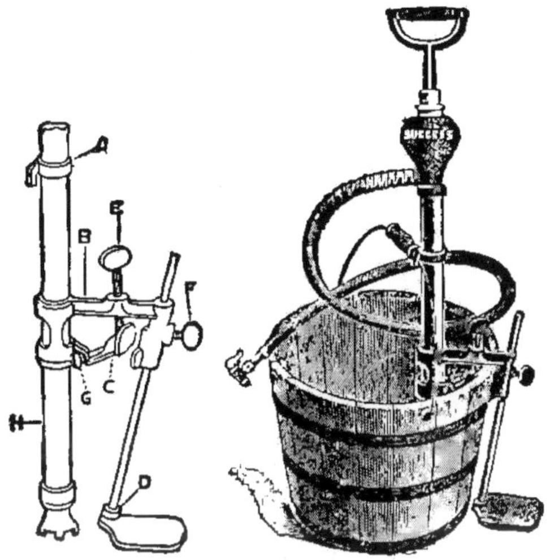

Fig. 69.

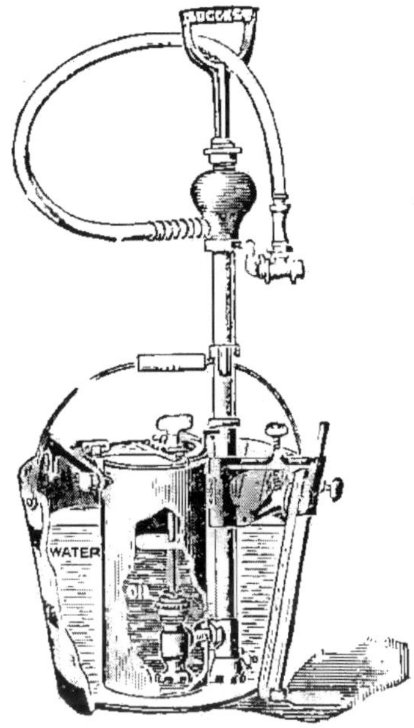

Fig. 70.

Fig. 69. A convenient and serviceable spray pump, using a common pail for a reservoir, to which it is attached by the device shown at the left.

Fig. 70. A similar pump with attachment by which kerosene and water may be sprayed together (294). Both are made by the Deming Co., Salem, Ohio.

the tree or plant, and with an atomizing nozzle (Figs. 69, 70, 71). For tall trees, the hose nozzle may be elevated by attaching it to the end of a light pole. In orchard spraying, the pump is often used on a wagon, and the man holding the hose sometimes stands on a high platform. A spray pump used on a large reservoir, needs an *agitator* to prevent the heavier part of the spraying mixture from settling.

Excellent bellows and force pumps, designed expressly for applying insecticides, are now manufactured.

305. The Use of Insecticides. In treating any given insect, the most important question to decide, is the manner in which it works its injury, as upon this will depend the preventive measures to be used.

306. Injurious Insects are referable to **Two Classes,** viz., *eating insects*, i. e., those feeding directly upon the plant tissues, as the potato beetle, the apple-tree borers,* the plum curculio; † and the *sucking insects*, i. e., those feeding only upon the juices of the plant, as plant lice, the squash-bug, ‡ and the oyster-shell bark-louse. §

307. The Eating Insects may be subdivided into *leaf-eaters*, those that devour the foliage: *root-eaters*, those that devour the roots: and *burrowers*, those that harbor within some part of the plant by eating a passage for their bodies.

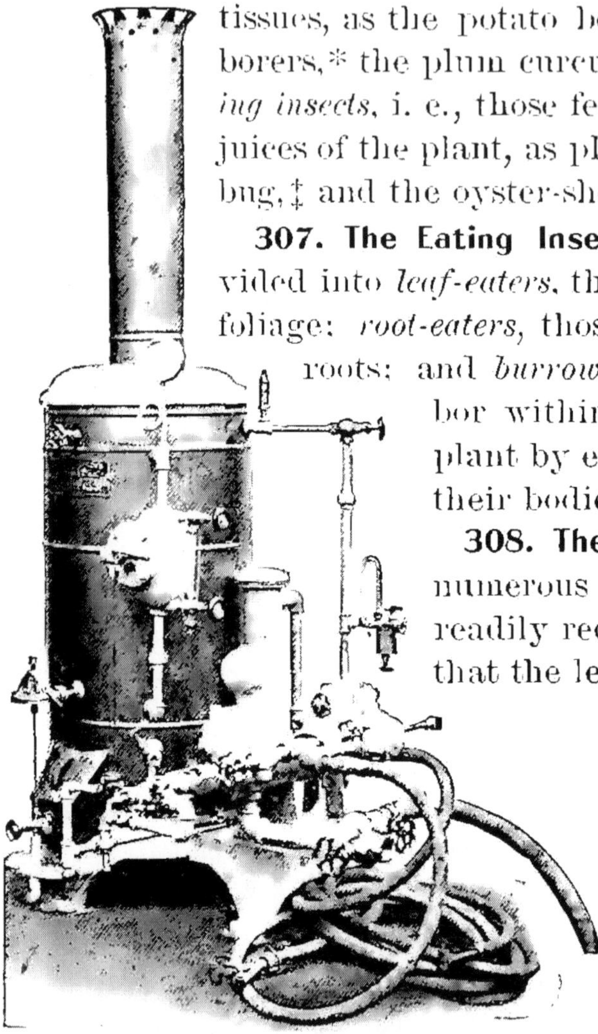

FIG. 71. Steam spraying outfit, manufactured by the Shipman Engine Co., Rochester, N. Y.

308. The Leaf-Eaters include numerous species. They are readily recognized by the fact that the leaves, on which they feed, disappear more or less rapidly. They may generally be destroyed by applying a poison to the foliage, for which purpose the arsenical compounds are well adapted (284). In cases where the use of a deadly poison is unsafe, hellebore (289) or pyrethrum (290) may be substituted.

* The round-headed apple-tree borer, *Saperda candida;* the flat-headed apple-tree borer, *Chrysobothris femorata.* † *Conotrachelus nenuphar.* ‡ *Anasa tristis.* § *Mytilaspis pomorum.*

309. The Root-Eaters include fewer species than the leaf-eaters and are usually more difficult to control.

Carbon bisulfid, injected into the soil about the roots of cabbage and cauliflower plants, with an instrument devised for the purpose (Fig. 72), has been successfully used to destroy the cabbage maggot,* and may be found useful in other cases. Attacks of this insect have also been successfully prevented by surrounding the stem of the young plant with small cards of thin tarred paper. One of these cards, the tool used for cutting them, and the manner of using the tool are shown in Figs. 72, 73 and 74.

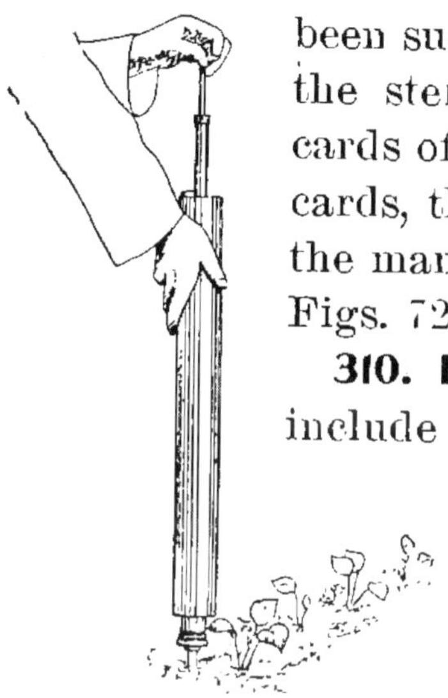

FIG. 72. Tool for injecting poisonous liquids about the roots of plants.

310. Burrowers, as the term is here used, include not only the so-called *borers* that burrow within the stems and roots of plants, and the *leaf-miners*, that live between the surface of leaves, but also the insects that pass their larval stage within fruits. Insects of this class are difficult to control, since they are mostly beyond the reach of insecticides.

311. Borers that infest the trunks and main branches of trees, may often be kept out by applying strong alkaline washes to these parts. Soft soap, reduced to the consistency of thick paste by a strong solution of washing soda, applied to the trunk or branches, forms a rather tenacious coating which repels the female insect. Painting the trunks of small apple trees a short distance above and below the surface of the ground with common paint

* *Phorbia brassicæ.*

or pine tar, is said to prevent the entrance of the round-headed borer (306). Protecting the trunk with straw or lath, as recommended to prevent sun-scald (186), may tend to keep out these insects. Borers in the trunk can often be destroyed by probing their holes with a flexible twig.

312. Leaf-Miners often infest spinach and beets grown for greens, rendering the leaves unfit for use. For these insects we can offer no preventive measures of established value. The application to the young foliage of powerful odorants, as coal-tar water or a solution of carbolic acid, may prove beneficial.

313. The Codling-Moth,* which causes so-called "wormy" apples and pears, is controlled by spraying the trees at the time of egg deposit, with water containing Paris green (285). The first spraying should be given as soon as the petals (143) fall, to be followed by a second six to ten days later. If much rain falls at this season, the sprayings may need frequent repetition. A drop of poisoned water should be lodged in the calyx (142) of every fruit, and as this evaporates, the poison deposited on the

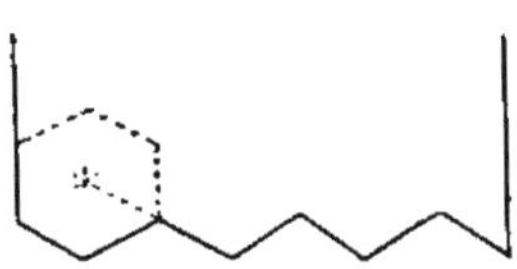

FIG. 75.

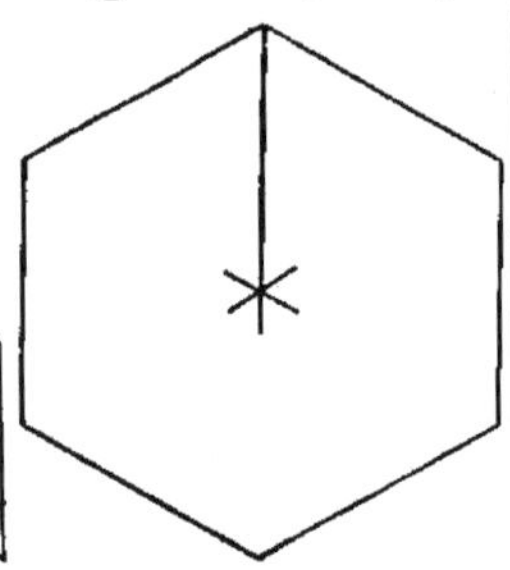

FIG. 73.

FIG. 74.

Fig. 73. Card of tarred paper, for placing about the stems of young cabbage and cauliflower plants. Reduced one-half.

Fig. 74. Tool for cutting the cards.

Fig. 75. Manner of using the tool. The dotted lines show the position of the edge of the tool on the paper.

* *Carpocapsa pomonella.*

skin kills the newly-hatched insect as it eats its way inward.

A band of cloth or paper, placed about the trunk of fruiting apple or pear trees forms a convenient retreat for larvæ of the codling-moth, in which to pupate. They may then be readily destroyed by removing the band. The bands should be a few inches wide, and should be put on before midsummer. They should be taken off once in ten to fourteen days, until the fruit is harvested, and all cocoons beneath them should be crushed.

314. The Plum Curculio (306) that so often stings young plums, causing them to drop before maturity, is controlled by jarring the beetles, that deposit their eggs in the young fruit, upon sheet-covered frames on cool, still mornings while their muscles are stiff (Fig. 76). The jarring should begin almost as soon as the petals (143) fall, and should be repeated every still morning as long as any beetles are found. The work must be done in the early morning. Any light wood frame, covered with cloth may be used as a substitute for the more convenient device shown in the figure. Where the substitute is used, the beetles must be looked for on the sheet and destroyed as found.

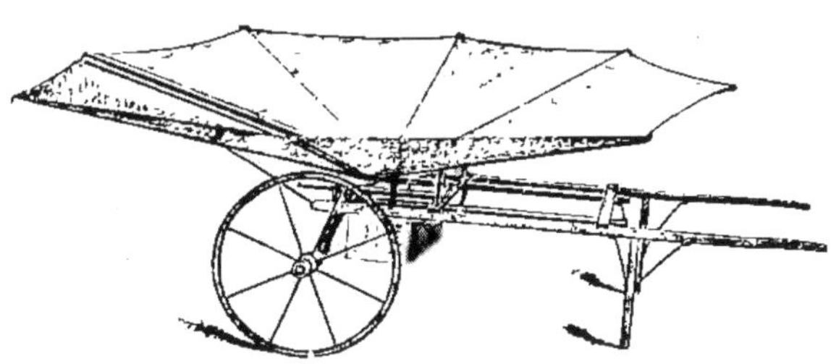

FIG. 76. Curculio catcher. It is wheeled beneath the branches of the tree, and the latter are struck with a light, cloth-covered mallet, which jars the beetles upon the sheet-covered frame, from which they roll into the box beneath. For small trees, the trunk slips in through the slot at the left.

315. The Prompt Destruction of **Infested Fruits** materially aids in keeping the fruit-burrowing insects in subjection. Hogs and sheep in the orchard are most valuable assistants in this work. The apple-maggot* is more effectually controlled in this manner than by any other known method.

316. Sucking Insects include many species. They feed on the juices of the plant which they infest, and do not directly devour its tissues, as do the eating insects; but they reduce its vitality by their continual drain upon the reserve food. The so-called *scale insects* belong to this class. These are especially difficult to destroy, since they are dormant the greater part of the year, and in this condition are protected by their comparatively resistant scales.

Sucking insects are not susceptible to poisonous insecticides, hence we must resort to materials that clog their breathing pores, as kerosene (294), that dissolve their eggs and scales, as potash solutions or that form an air-tight coating over them, as the resin washes (295).†

317. The Life Histories of **Injurious Insects,** which can not here be taken up, may profitably be studied by the plant grower. A standard work on economic entomology will furnish the needed information.

B—PLANTS AS AFFECTED BY VEGETABLE PARASITES.

318. Many of the most serious enemies of cultivated plants belong to this class. As a rule, vegetable parasites contain no chlorophyll, and hence are incapable of forming their own food. While most of them belong to the lower orders of plants, a few species are highly developed and produce true flowers and seeds.

---

* *Trypeta pomonella.*

† The cottony cushion scale, *Icerya purchasi,* which was very destructive to the orange in California, has been nearly suppressed by the introduction of an Australian parasite, the *Vedalia cardinalis.*

a By Flowering or phanerogamic (phan′-er-o-ga′-mic) **parasites.**

Of these, the only ones sufficiently common or injurious to need mention are the broom rapes and the dodders.

319. The Broom Rape of **Hemp** and **Tobacco,*** is the most injurious species of this class. The seeds germinate in the soil, and the young plants attach themselves to the roots of their host which they enfeeble by robbing them of nourishment. In the case of hemp, the parasite also injures the quality of the fibre.

Preventives. The seeds of hemp or tobacco should not be taken from a crop infested with broom rape. Infested fields should be planted for several years to some crop not attacked by broom rape, as potatoes, Indian corn, beans, grains or grasses. In infested crops, the broom rape should not be permitted to mature its seeds.

320. The Dodders of **Clover** and **Flax,**† are the most injurious of their class. The young plant attaches itself to the stem of its host, about which it twines, robbing it of nourishment by means of small suckers.

Preventives. The seeds of dodder are somewhat smaller than those of clover or flax, and hence may be separated from the latter by sifting. Badly infested ground should be devoted for two to four years to a crop not attacked by the dodder.

b — Plants as affected by fungous parasites.

321. The Fungi constitute an extensive class of plants that derive their nourishment wholly from organic matter. Many of them are injurious to cultivated plants. Unlike the harmful insects, most of which work their ravages

---

* *Philipœa ramona.* † *Cuscuta trifolia, C. Epilinum.*

within full view, the fungi are in many cases discernible only with the microscope, and reveal their presence only by the death or injury of their host. The fungous parasites are very numerous and exhibit great diversity of structure and habit. Some of them live only upon enfeebled plants, while others attack healthy ones. Some, as the pea mildew, grow upon the surface of their host, drawing their nourishment through the epidermis; others, like the peach curl and oat smut, grow within the tissues of the plant upon which they feed. All of the latter class send their fruiting parts to the surface of the host plants to disseminate their spores in the open air.

The fungi multiply from extremely minute spores (53) that are produced in immense numbers, and when mature, are very readily blown about by wind. Many of them also multiply from thread-like organs called *hyphæ* (hy′-phæ), something in the same manner as Canada thistles multiply from their roots.

322. Methods of **Controlling Fungi** are of three classes:

a — Removing and destroying the affected parts;

b — Preventing the germination of the spores;

c — Destroying the fungus itself by applying some destructive material (a *fungicide* (fun′-gi-cide.)).

323. Destruction of the **Affected Parts** is the most effectual preventive known in cases where the fungous disease attacks a portion of the plant whence it spreads to the remaining parts, as in the black knot of the plum,* the blight of the pear, apple and quince,† the red rust of the raspberry and blackberry,‡ and the corn smut.§

* *Plowrightia morbosa.* † *Micrococcus amylovorus.*
‡ *Cæoma luminatum.* § *Ustilago Maydis.*

The affected part should be removed as soon as discovered and burned at once, to destroy any spores of the fungus it may contain or which might mature later. It is generally important to cut the diseased branch some distance below the point of visible infection, as in many cases the mycelia of the fungus extend farther than external appearances indicate.

324. Preventing Spore Germination is the only known method by which we can combat the fungi developing *within* the host plant (*endophytic* (en-do-phyt′-ic) *fungi*).

In fungi that develop from spores planted with the seed, as the smuts of the small grains, spore germination may be prevented by treating the seed with a solution of certain chemicals or with hot water. Of the former, sulfate of copper (copper sulfate, blue vitriol, bluestone) has been most used, and unquestionably destroys the spores of the smuts, but it has generally been found to injure more or less the germination of the seed.

325. The Hot-Water Treatment has proven fully as successful in preventing smut as the preceding method, without injuring the seed. This treatment consists in immersing the seed for ten minutes in water at a temperature of about 135° F. For treating a quantity of seed, some special provisions are necessary, as it is somewhat difficult to bring every seed in contact with the water at the proper temperature. Provide two large vessels, as two kettles over a fire, or two boilers over a cook stove — one to contain warm water (110–120° F.), the other to contain hot water (132–135° F.). Place a reliable thermometer in the hot-water vessel that the temperature may be watched. The warm water is used to warm the seed, before dipping it in the hot water; otherwise it is difficult to maintain the temperature of the latter. The seed is placed in a covered basket, preferably of wire cloth, in quantity not exceeding one-eighth of the volume of the water; the basket should be but partially filled. Immerse the basket several times in the warm water, a moment at a time, giving it a rotary motion in order to bring every seed in contact with the water. Then plunge it into the hot water and repeat the immersions as before, carefully watching the thermometer in the meantime. Should the temperature fall below 132°, add water of a higher temperature; and if it rises above 135°, add cold water. After the seed has been in the hot water ten minutes, remove the basket and plunge it into cold water, then spread the seed out to dry. The drying need not be thorough unless the seed is to be stored for a time.

326. Fungi that **Develop** from **Spores Surviving** the **Winter In** or **Upon** the **Soil,** as the onion* smut, cannot be prevented by disinfecting the seed. For this disease a mixture of flowers of sulfur and air-slacked lime, sown with the seed, has proved beneficial by preventing infection of the young plant.

327. Fungi the **Spores** of which **Survive** the **Winter Within** their **Dead-Host Plants,** as in the club-root of the cabbage† and turnip, and the onion mildew,‡ may be held in check to some extent by burning the fungus-killed plants at the close of the season.

328. Fungi that **Infect** their **Host** from **Spores Deposited On** the **Aerial Parts** of the plant, as the scab of the apple §

* *Eurocystis Cepulæ.* † *Plasmidiophora Brassicæ.*
‡ *Peronospora Schleideniana.* § *Fusicladium dendriticum.*

and pear, and the downy grape-vine mildew * may be held in check by applying a fungicide (321) to the host plant, to destroy the spores as they alight upon it. Various compounds of copper and of sulfur are destructive to the spores of fungi, and when properly applied, are harmless to the plant. The copper compounds are more generally satisfactory, since they have the greater adhesive power.

FIG. 77. Apple affected with scab (the dark spots), *Fusicladium dendriticum*. (After Scribner).

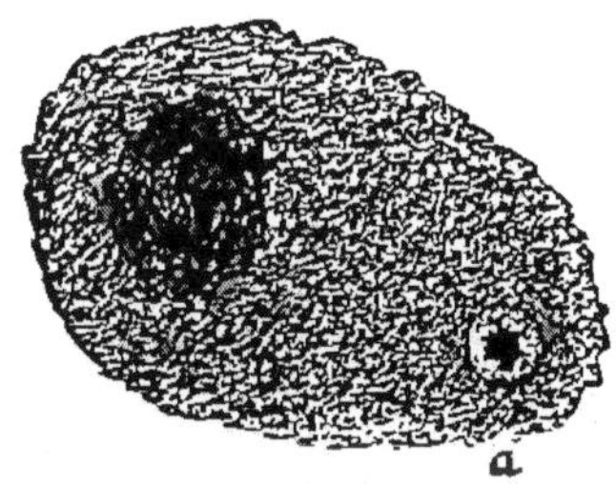

FIG. 78. A scab spot magnified. (After Trelease).

FIG. 79. Section through a scab spot, highly magnified. The egg-shaped parts at the right are the spores. (After Trelease).

329. The Bordeaux Mixture, which consists of a compound of copper sulfate (324) and lime, is now extensively used to prevent many fungous diseases of this class. A standard formula for the Bordeaux mixture is:

Dissolve 6 pounds of copper sulfate in 4 gallons of hot water; in another vessel slack 4 pounds of fresh quicklime in 4 gallons of (hot or cold) water. When both are cool, pour the contents of the two vessels together and

* *Peronospora viticola.*

add enough water to make 45 gallons of the whole. Metal vessels, other than those of brass or copper, should not be used.

The hot water hastens the dissolving of the copper sulfate. Cold water may be used by suspending the sulfate in it in a sack of coarse texture a day or two in advance.

Prepared by the above formula, the Bordeaux mixture often contains more lime than is needed for the chemical action that occurs. To avoid this excess of lime, a chemical test may be used, as follows: Pour only half of the slacked lime and water into the copper-sulfate solution, stir well, and add a few drops of a 20 per cent solution of potassium ferrocyanid. If a rich, reddish-brown color is produced, add more lime. Continue to test and add lime until the reddish-brown color no longer appears. Then add a little more lime, as *a slight excess of lime is desirable.* A bright, clean knife blade may also be used as a test. If a slight film of copper forms upon it when placed in the mixture, more lime is needed. The Bordeaux mixture is preferably strained before use, and should be kept well stirred during its application. It may be applied with any good spray pump.

The arsenical compounds (284) may be added to the Bordeaux mixture, and thus a single treatment will serve for both insects and fungi.

330. The Diseases Preventable by **Bordeaux Mixture** are the apple and pear scab (328), the downy mildew and black rot * of the grape, the early † and late blight ‡ of the potato, the gooseberry mildew,* the leaf-blight of the pear,† and some others.

* *Laestadia Bidwellii.* † *Macrosporium Solani.* ‡ *Phytophthora infestans.*

In all these diseases, however, *the treatment is preventive rather than curative.* The first application should be made before the disease appears and should be followed occasionally by others as new foliage is formed or as the material is washed off by rains.

331. Ammoniacal Solution of **Copper Carbonate** possesses nearly the same fungicidal properties as Bordeaux mixture, but adheres less strongly to foliage. Being a solution, it requires no straining or stirring, and it leaves less stain on drying than Bordeaux mixture, which makes it preferable to the latter for use upon plants of which the fruit is nearly mature. To make this solution, dissolve one and a half ounces of precipitated copper carbonate in one quart of strong commercial ammonia, and add 25 gallons of water. The ammonia should be procured in a glass or earthen vessel, which should be kept tightly corked. To prevent waste of the ammonia by evaporation, prepare immediately before spraying.

332. Potassium-Sulfid Solution is used to some extent to prevent gooseberry mildew (330), and a few other diseases, but it is less enduring in its effects than the copper compounds. To prepare it, dissolve one-half ounce of potassium sulfid (sulfuret of potassium, liver of sulfur) in one gallon of water and apply immediately. The sulfid is best dissolved in a little warm water and then diluted.

333. Moisture Favors Spore Germination, hence a free circulation of air through the orchard and vineyard tends to prevent fungous diseases by absorbing excessive moisture (227). Branches of fruit trees should not be per-

* *Sphærotheca Mors-uvæ.* † *Entomosporium maculatum.*

mitted to hang near the ground, and weeds should be kept down.

Grapes are sometimes inclosed in paper bags on the vine, to keep them dry, and thus preserve them from fungous attack. Grape vines sheltered from rains by a cornice are seldom much troubled with fungous diseases.

334. Fungi that **Develop** chiefly on the **Outside** of the **Plant** (*epiphytic* (ep-i-phyt′-ic) fungi), are as a rule readily controlled by sulfur, either in the form of flowers of sulfur, or the solution of potassium sulfid (332). To this class belong the powdery mildews of the grape,* apple,† etc.

335. The Cultivator will often **Need** to **Consult** the **Specialist** in dealing with fungous diseases. In many cases, it will be difficult or impossible for him to decide as to the exact nature of a given trouble without careful training, and skill in the use of the compound microscope. Specialists in this line are now employed by the governments of most civilized nations and by many agricultural experiment stations, and they should be freely consulted. Much may be learned, however, by studying the best books on the subject. The cultivator should be able to recognize the principal fungous diseases.

SECTION VIII. PLANTS AS AFFECTED BY WEEDS

336. Weeds are plants of the higher orders that persist in growing where they are not wanted. They injure the desirable plants about which they grow, by robbing them of light, moisture and food, and their presence is an evidence of slovenly culture. The remarkable vigor and prolificacy possessed by many weeds would enable them

* *Uncinula spiralis.* † *Podosphæra oxycanthæ.*

to soon overcome most cultivated plants, but for the aid of the cultivator. As with harmful insects and fungi, prompt and persistent efforts are essential to the control of weeds in most cultivated grounds.

337. Annual, Biennial and **Perennial Weeds.** With reference to their term of life, weeds and other plants are divisible into three classes, viz., *annual*, those that live but one season; *biennial*, those that live only two seasons; and *perennial*, those that live an indefinite number of seasons. Weeds of the first class usually seed most abundantly, and hence they are most widely distributed and appear in cultivated grounds in the greatest numbers. Those of the third class are commonly most tenacious of life and are therefore often most difficult to control.

FIG. 80. Showing how plants of the sow thistle multiply from underground stems.

338. *Annual* and *biennial weeds*, since they have a definite life period and multiply almost exclusively by seed, may be controlled by preventing seedage. To accomplish this with certainty, the plants should be destroyed before bloom, as many species possess enough reserve food to mature seeds sufficiently for germination, if cut while in flower.

339. *Perennial weeds* often multiply by suckers as well as by seeds (Fig. 80). Since the roots or underground stems whence the suckers grow (115), are hidden beneath the soil and are often extremely tenacious of life, weeds of this class are frequently very hard to eradicate. Persistent prevention of leafage, by starving the protoplasm of the roots, is always effectual, though it is often very difficult to carry out, since the suckers of some species grow with great rapidity. Yet, on the whole, no better remedy is known. Frequent plowing and cultivation of the infested ground is usually the most effectual means of preventing leafage.

Certain very tenacious perennial weeds, as the Canada thistle * and the sow thistle, † when growing on deep, rich loams in which the roots spread freely below the plow line, may, it is said, be crowded out by seeding the land to grass, at less cost than they can be subdued by the plow.

If we have mastered the foregoing chapters, we are now prepared to enter upon a more advanced stage of culture, and to learn how to cause new plants to grow, and how to so treat the plants thus grown that they may best serve our purpose.

The following books are recommended for reading in connection with the preceding chapter: Elementary Meteorology, Waldo; Chemistry of the Farm, Warington; The Spraying of Plants, Lodeman; Economic Entomology, Smith; Fungous Diseases of the Grape and Other Plants, Lamson–Scribner; American Weeds and Useful Plants, Darlington.

* *Cnicus arvensis.* † *Sonchus arvensis.*

CHAPTER IV. PLANT MANIPULATION

SECTION I. PLANT PROPAGATION

340. Propagation, as the term is generally used in plant culture, is the artificial multiplication of plants, i. e., reproduction (16) encouraged or induced by the knowledge, skill and care of the cultivator.

Theoretically, any part of a plant containing living protoplasm, with sufficient prepared food or tissue capable of preparing food (59), may under proper conditions develop into a complete plant. But in practice, we have not been able to fully demonstrate this theory; for example, the roots and leaves of some plants have not been induced to form buds.

341. Plants are **Propagated** by **Numerous Methods,** but only two of these are distinct in kind, viz., *by seeds (or spores)*, and *by division of the plant.* In propagation by seeds, the *embryo of the seed* (54) is the vital center whence the new plant is developed. In propagation by division, *a living bud* (128) from the parent plant, or a bit of tissue capable of forming a bud, is substituted for the embryo of the seed. In seed propagation, the resulting plant is the product of sexual fecundation (150), and hence cannot be considered as strictly a part of the parent. It does not necessarily resemble the parent closely. In propagation by division, on the other hand, the resulting plant may be regarded as simply a continuation of the growth of the parent in a new location, and generally closely resembles the parent.

342. Propagation by **Seeds** is commonly practiced with annual and biennial plants and with perennials in which the reproduction of the exact parental form is unimportant, as in the cereals, forest trees and seedlings intended for grafting. This method is also used when variation in the progeny is desired, as in developing new varieties (438 b).

343. Propagation by **Division** of the plant is used when it is desired to reproduce the exact parental form, as in fruit- and the finer ornamental trees, many flowering plants, etc.; in certain plants that are more readily multiplied by division than by seeds, as mint and many other perennial herbs; and in other plants that rarely or never produce seed, as the horse-radish, sugar cane, banana, etc.

A — Propagation by Seeds

344. This is the most common method of propagating plants. It seemed appropriate to give nearly all of the needed directions for planting seeds in the first two sections of Chapter II. We add, therefore, only a few general rules deduced from the principles there stated.

a — *The soil in which seeds are to be planted should be thoroughly crumbled*, because the seeds must have access to the oxygen of the air, or they cannot germinate (31).

b — *The well-crumbled soil should be compactly pressed about the seeds*, because the seeds cannot absorb moisture rapidly unless the seed-case is in contact with the moist soil particles at many points (27 b).

c — *The soil should be moist, but not wet enough to puddle* (31); otherwise the oxygen is likely to be shut out from access to the seeds (35).

d — *Seeds should be planted no deeper than is necessary to insure the proper degree of moisture;* otherwise the plantlet expends a needless amount of energy in reaching the surface (51, 48). Very small seeds should be only slightly covered, if at all, and must receive artificial watering when necessary (52). Spores must not be covered with soil at all (53).

B — Propagation by Division.

345. We have seen that a part of a plant, placed under favorable conditions, is usually capable of developing into a complete plant (340). A section or cutting of the *stem*, for example, that has no roots at the time it is cut off, may be caused to form roots, and thus become a complete plant. A cutting of a *root* may also put forth a bud, which in turn may develop into a shoot, and form leaves, flowers and fruit. Again, we have seen that portions of cambium from different, nearly-related plants may unite by growth (70), which enables us to change undesirable sorts into valuable ones by grafting (383). These and certain other methods of multiplying plants, come under propagation by division.

In propagation by division, the presence of at least one healthy growing point (67) *in the part selected for the propagation is generally essential to success* and is always helpful.

The processes treated in this and the two succeeding sections may be likened to surgical operations in medicine. If plants are less highly organized and possess less of sensibility than the higher animals, they are, none the less, living beings. Violent operations, if necessary, should always be performed with this truth in mind. *Needless injury and careless handling in the treatment of plants are always to be avoided.*

346. Two Methods of Propagation by Division may be distinguished, viz., by *parts intact* and by *detached parts.* In the first, the part selected for propagation is not separated from the parent until the organs needed to make it self-supporting are formed; or if a cion (386), until it has united to the part on which it is intended to grow. In the second method, the part intended for propagation is severed from the parent at the outset and placed under conditions favoring the formation of the organs needed to make it self-supporting; or if a cion, favoring its union with the stock (383).

A—PROPAGATION BY PARTS INTACT.

This method is applicable to many plants and has the advantages of being reliable and requiring little skill. The part selected for propagation, being nourished by the parent until it forms the needful organs, is able to endure unfavorable conditions that would prove fatal in most other methods of propagation. This method includes four divisions, viz., propagation by *suckers* (347), by *stolons* (348), by *layers* (349), and by *approach grafting* (399). In the first two, the propagation is performed by the parent plant without other aid than the maintenance of a well-aerated, moist and clean soil that stimulates the production of the needed organs, which in these cases are roots.

347. Propagation by Suckers. Suckers are shoots that originate from roots or underground stems and grow upward, forming young plants about the parent, as in the blackberry, plum, choke-cherry, etc. The propagation consists in simply cutting off the root or underground stem whence the sucker proceeds, and transplanting the latter.

The growth of suckers may generally be stimulated in plants that naturally produce them, by cutting off the roots or underground stems from which they grow, or by severely pruning the top.

The propagation of woody plants from suckers is not, as a rule, considered wise, since the roots are usually poorly developed in proportion to the stem, and some plants grown in this manner seem to acquire the tendency

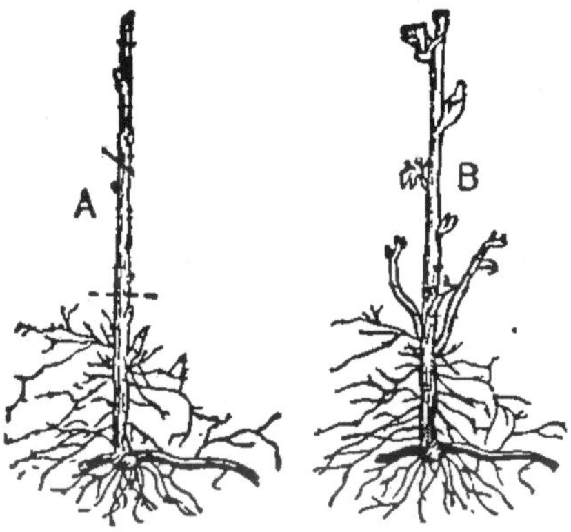

FIG. 81.

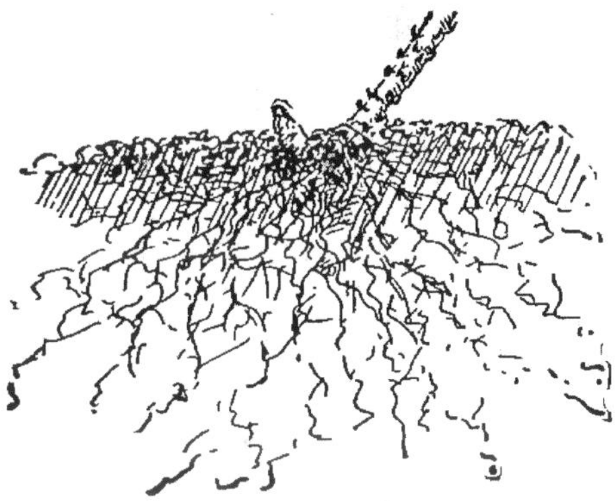

FIG. 82.

Fig. 81. Sucker plants of the red raspberry, *Rubus strigosus*. A, before growth has started; B, after. The two shoots of B starting just above the roots form the new canes.

Fig. 82. Tip plant of black raspberry. The bud, whence the young shoot starts, appears at the base of the parent cane. (After Bailey).

to form suckers excessively. In the red raspberry * and the blackberry, † however, propagation by suckers is the most convenient method, and it appears to be followed by no bad results (Fig. 81).

348. Propagation by **Stolons.** A stolon is a branch that starts above or at the surface of the ground and either grows prostrate or curves downward till it reaches

* *Rubus strigosus, R. Idæus.* † *R. villosus.*

the ground, where it takes root, usually at the nodes (116). The currant, juneberry, cranberry and many herbaceous plants are readily multiplied in this way. Stolons often root without assistance, but the rooting is much hastened and encouraged by covering the branch with soil. When well rooted, the young plants may be separated from the parent by cutting the stolons.

Woody plants grown from stolons are seldom uniform in size and are not often as well rooted as those grown from cuttings (358). Some herbaceous plants are, however, more readily propagated by stolons than by any other means.

The *offset* by which the houseleek * is so readily propagated, is a very short stolon that forms a single tuft of leaves at its apex. The cane of the black-cap raspberry, † which roots from the tip (Fig. 82), and the runner of the strawberry (Fig. 83), that forms a plant at each alternate node, are modified stolons.

349. Propagation by **Layers** or **Layering.** The layer is an artificial stolon, i. e., a branch that does not naturally grow downward, which is covered with or surrounded by moist soil to stimulate the production of roots (89). The branch may be bent down and covered, as is usually practiced with the grape, wisteria etc., or the soil may be ridged up about the branch, as is done with

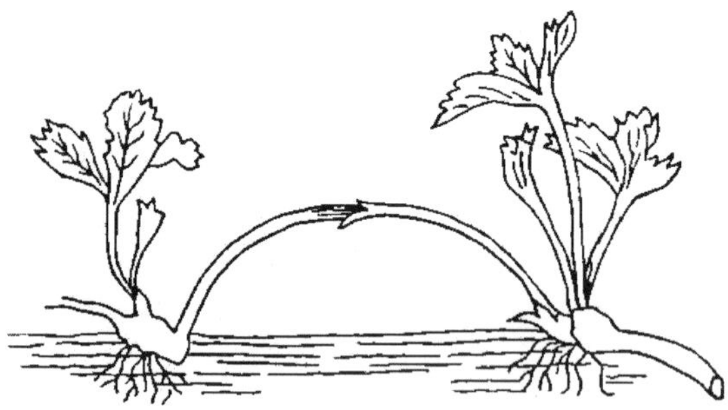

FIG. 83. Runner of the strawberry.

* *Sempervivum.* † *Rubus occidentalis.*

the quince and paradise apple. In either case, the terminal portion of the stem is commonly left uncovered. In the latter method, which is known as *mound-layering* (Fig. 84), the stems of the plant to be layered are usually cut off just above the surface of the ground in early spring, to stimulate the formation of vigorous shoots, which are ridged up about midsummer or preferably not until the succeeding fall or spring. The ridging should be sufficiently high to cover several of the lower nodes (116). Roots grow out at the nodes and the shoots are usually well rooted by the autumn following the ridging.

FIG. 84. Mound-layering of gooseberry plants. (After Bailey).

Many woody plants that do not readily form roots when layered, may be induced to do so by mutilating the stem somewhat in the covered part. This tends to restrict the growth current (80) and causes an accumulation of reserve food, from which roots may grow. Girdling, twisting, bending or spliting the stem for a short distance will often have the desired effect (Fig. 85).

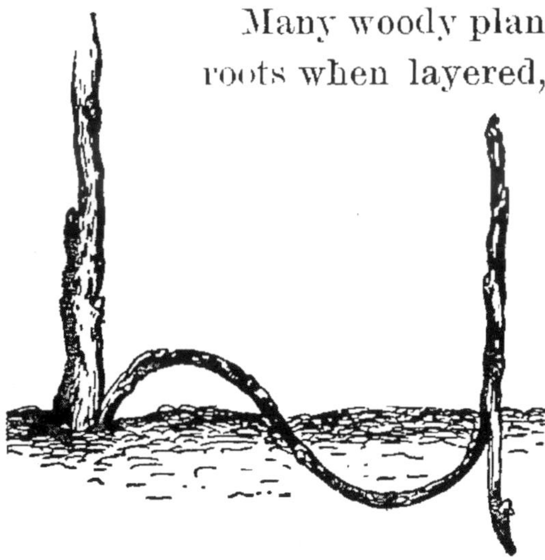

FIG. 85. Layered branch of currant, split to encourage the formation of roots.

Layering is a very reliable and expeditious method of propagating many woody and herbaceous plants.

350. Propagation by **Division** of the **Crown** of the plant, which is practicable with many perennial herbs, as the rhubarb, dahlia, globe artichoke etc., though not strictly analogous to propagation by stolons or layers, may be considered here. It consists in taking up the plant, preferably while dormant, and cutting the crown into two or more parts, according to its size or the number of plants desired, and planting the divisions as separate plants. This method is applicable to propagation for private use, rather than for sale purposes.

Propagation by *approach grafting*, although in order here, is more readily treated with the other methods of grafting (399).

B—PROPAGATION BY DETACHED PARTS.

This comprises two different modes of propagation, viz., by *specialized buds* and by *sections of the plant.*

a—Propagation by Specialized Buds.

351. This includes propagation by *bulbs, bulblets, corms* and *tubers.* It is in a sense intermediate between propagation by parts intact (346) and by cuttings (358). The bud that is to form the future plant, though not having roots of its own, has been specially prepared by the parent, through an abundant food supply and a partially dormant condition of the protoplasm, to maintain a separate existence, even under adverse conditions, and in due time to develop into a plant. In these respects it resembles the seed, from which it differs, however, in the less dormant condition of its protoplasm and in not being the product of sexual fecundation (341).

352. The Bulb is a very short stem containing a terminal bud inclosed in scales (128). The scales are thickened by a store of food, and in their axils are smaller lateral buds. The terminal bud usually develops into a flower, and then perishes. One or more of the lateral buds may develop into flower-buds for the next year, and thus continue the life of the plant, as in the common onion (Fig. 86); or the lateral buds may develop at the expense of the parent, as in the potato onion.

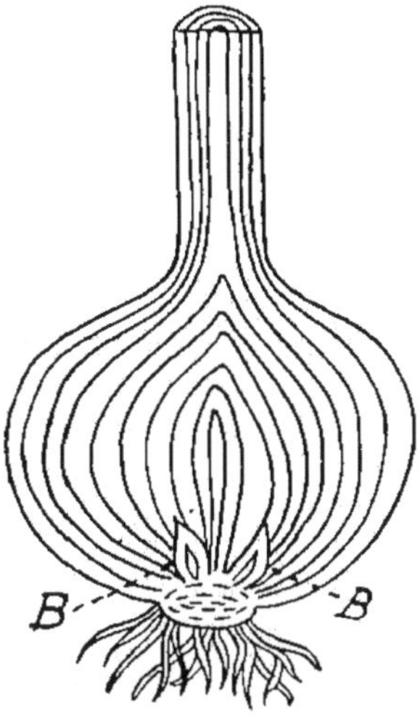

FIG. 86.

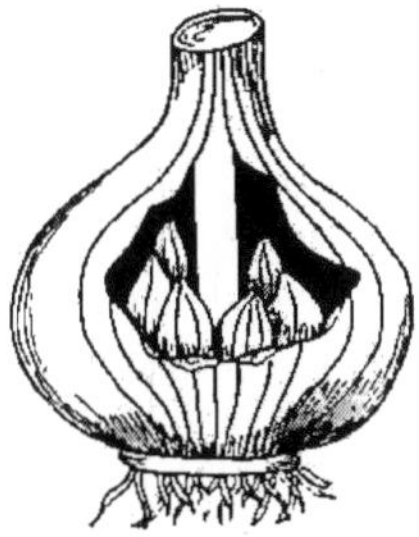

FIG. 87.

FIG. 88.

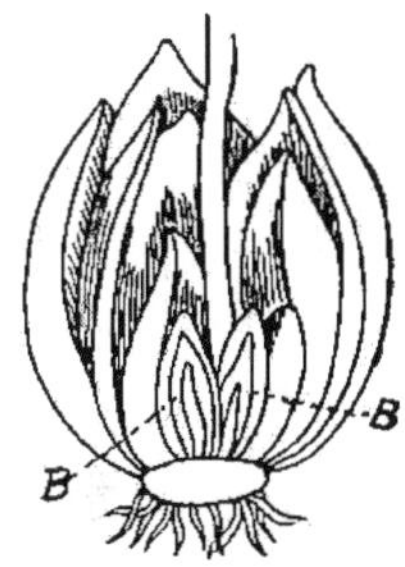

FIG. 89.

Fig. 86. Bulb of the common onion, *Allium cepa*, divided lengthwise. B, buds.

Fig. 87. Bulb of garlic, *Allium sativum*. It contains several smaller bulbs (cloves).

Fig. 88. Bulb of wild lily.

Fig. 89. The same divided lengthwise, showing buds, B.

353. Bulblets or **Bulbels** are small bulbs formed in the axils of the leaves in certain plants, as the tiger lily,* (Fig. 90), or at the apex of the stem, as in the "top" or bulb-bearing onion (Fig. 91).

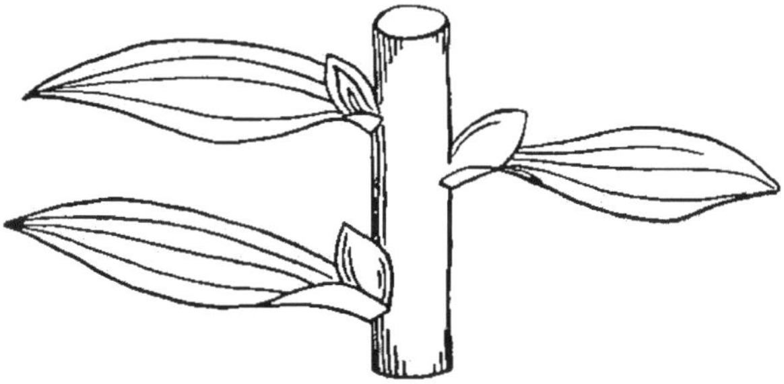

FIG. 90. Bulblets in axils of leaves of tiger lily.

* *Lilium tigrinum.*

354. The Corm (Fig. 92) differs from the bulb chiefly in being without scales. The food is deposited in the thickened stem. The corms of our flowering plants, as the crocus, cyclamen etc., are generally called bulbs in commerce.

FIG. 91. Bulblets of "top" onion, sometimes used as onion "sets."

FIG. 92. Corm of crocus, with small corms (buds) for following year.

355. The Tuber, of which the common potato is the most familiar example, differs from the corm in being the end of an underground branch of the stem (115), instead of developing in direct contact with the parent. It also has more numerous buds (eyes) than the corm.

356. Propagation from **Bulbs, Bulblets, Corms** and **Tubers** is a very simple operation and consists merely in planting these parts in the place where the plants are desired. Tubers may be cut into pieces containing one or more buds each, if desired. The rules given for planting seeds (344) apply equally well here. All should be stored for preservation in a cool, moderately dry place, that is free from frost. They retain their vitality but a single year.

In the methods of propagation thus far considered, with the sole exception of layering (349), advantage has been taken of a natural mode of plant multiplication. The skill of the cultivator, however much it may assist the processes, is not necessary to their success, since wild plants habitually increase by the same methods. We will now consider a method which is less often illustrated in nature, and in which the skill and care of the cultivator are, as a rule, essential to its accomplishment, viz.:

b — Propagation by sections of the plant.

The various methods of propagation in this division are alike in the fact that a detached part of the parent plant, containing living protoplasm, is placed for a time under specially favorable conditions, in virtue of which the part is enabled not only to live, but to perform its functions and reproduce the needed organs; or if a cion (386), to unite by growth to the part with which it is placed in contact.

357. In propagation by sections of the plant we must, of necessity, wound the plant tissues in securing the parts for propagation. Since it is always desirable that the wound should heal promptly (73), *it is important that the cutting tools used should have sharp and smooth edges.*

As here considered, propagation by sections of the plant includes two methods, differing materially in their requirements and in the manner of development of the plants, viz., propagation by *cuttings* and by *grafting.*

a — *Propagation by Cuttings.*

358. A Cutting is a detached member of a plant, intended to be placed in the soil or some other medium for propagation. It may be in an *active* or a *dormant* state (13), and may or may not contain a growing point (67). Before the cutting can become a plant, it must develop the essential part or parts of the plant that it lacks: i. e., the stem and the leaves, or the root, or all these members. Cuttings of the stem are usually planted with their proximal end (116) in the soil, and their distal end in the air. Root cuttings are generally covered in the soil.

359. Nearly All Plants may be **Propagated** by **Cuttings** from one or another of their parts. The ease with which plants may be multiplied in this way varies greatly in different species (21), and even in different varieties of the same species. The appearance of a plant does not always indicate the facility with which it may be grown from cuttings; the only sure way to ascertain this is by trial.

Climate exerts a marked influence upon the tendency of plants to develop from cuttings. In certain locations in southern Europe and in parts of South America, branches of the common apple tree, sharpened and driven into the ground as stakes, often take root and sometimes even bear fruit during the same season. A warm, moist atmosphere is very favorable to propagation by cuttings.

We have seen that the roots of certain plants normally develop buds (131). In like manner, the stems of many plants, as the potato, grape etc., normally develop growing points of roots at their nodes (116). *Plants that normally develop buds upon their roots, or growing points of roots at their nodes, are readily propagated by cuttings.* But propagation by cuttings is not limited to such plants (362).

360. The Essential Characteristics of a **Cutting** are *a*—a certain amount of healthy tissue; *b*—a certain amount of prepared food, or of tissue capable of preparing food (59); *c*—in most species, a growing point (67), either of the stem or root, or of both.

361. The Parts of plants to be **Used** for **Cuttings,** therefore, are preferably the younger, matured growths, since the tissues of these are most vigorous: or else a part that possesses a certain amount of healthy and vigorous leaf tissue. The cutting should always contain one or more buds when practicable (128).

362. Conditions that **Favor** the **Growth** of **Cuttings.**

a — *A soil warmer than the air above it* ("bottom heat") is important in growing many plants from cuttings. Warmth stimulates plant growth, and when applied to one part of a plant, it stimulates growth in that part. If the soil about a planted cutting is warmed to a temperature considerably higher than that of the air above, the growth of roots is stimulated. Indeed bottom heat often excites growth in cuttings that will not grow without it.

b — *A comparatively low air temperature* is important in growing many plants from cuttings of the stem (377), because it is essential that the stem growth be held in check until roots are formed. A soil temperature of about 65° F., with an air temperature about fifteen degrees lower, is suited to the great majority of plants usually propagated under glass from cuttings. It is important that these temperatures be maintained nearly constant until roots have developed.

Since we have better facilities for raising than for lowering the natural temperature of the atmosphere, propagation from cuttings is easiest at a time of the year when the temperature of the atmosphere during the day does not much exceed 50°. By observing special precautions, however, it is possible to propagate many plants from cuttings during the warm season.

c — *Abundant moisture* is important in growing plants from cuttings, because moisture favors root development (89), and water is essential to cell growth (63). The amount of water required varies considerably with different plants and conditions.

With cuttings containing leaf tissue (377, 382), transpiration (75) must be reduced to the minimum until roots are formed, because water cannot be taken up freely without root-hairs (101). For such cuttings, therefore, the air as well as the soil must be kept abundantly moist (369), and the direct rays of the sun must be intercepted by shading (236).

363. Methods for **Controlling Temperature.** The alternations of temperature in the open air are unfavorable to the development of cuttings, though many plants, as the willow, grape and currant, are readily propagated from cuttings out of doors. Some structure, therefore, that may confine warmth radiated from the earth or artificially generated, or that may when necessary shut out a part of the solar heat, is always of great assistance in propagating plants from cuttings, and in many species is essential to success. Since light is necessary to food preparation (59), such a structure must be roofed with glass or some other more or less transparent material.

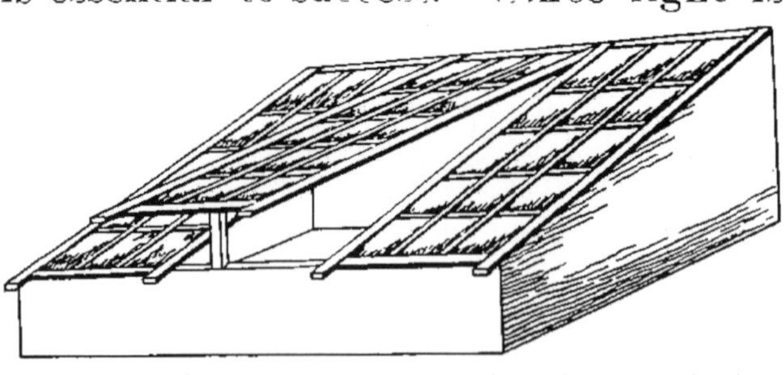

FIG. 93. Cold-frame, with sash lifted for ventilation.

364. The Cold-Frame (Fig. 93) is the simplest structure of this kind. It consists of a frame or box without bottom, usually shallower on one side than on the other, covered with glazed window sash.* The frame is generally placed so that its shallower side faces the south, thus giving its cover a southward slope. It has no provision for artificial heat, though when covered with glass, the temperature within the frame is much increased during sunshine, owing to the property possessed by glass of confining the heat rays. The cold frame should be protected in freezing weather by an additional cover of mats or blankets, while excessive sun heat should be avoided by shading (236). Muslin- or paper-covered frames require no shading.

Although affording no bottom heat (362 a), the cold-frame may be used for propagating many plants from cuttings. It is also serviceable in connection with the propagating bed (368) for "hardening off" young plants grown from cuttings in the latter, as well as for growing many plants from seed. Set over a pit in the earth, the cold-frame makes an excellent place (cold pit) for wintering half-hardy plants.

365. The Hotbed differs from the cold-frame in having bottom heat (362 a), which is usually supplied by the fermentation of moist vegetable material, as horse manure, leaves, refuse hops or tan-bark. The material intended for heating, if fresh, should be thrown into a pile of sufficient size to generate heat several days before it is desired for use; and unless already moist, it should be moderately sprinkled with water. In order that all the material may reach the same stage of fermentation, the mass should be made into a new pile after the heating

* Muslin or paper is sometimes used instead of glass, and these materials may be rendered waterproof and less opaque by painting with linseed oil or some similar material.

starts vigorously, as is indicated by vapor rising from the heap, and the outer part of the mass should be placed in the center of the new pile. Leaves ferment slower than the other materials above named, and hence may often be advantageously mixed with them to lengthen the period of fermentation.

Heat is economized by placing the fermenting material in a pit in the ground, but hotbeds are often made above ground. The hotbed pit should be in a well-drained and sheltered place, and two to two and one-half feet deep. In this the heating material should be moderately packed, until the pit is nearly or quite full. The frame may then be placed over the pit, after which the heating material should be covered with soil and the sash put on to confine the warmth. Within a few days after covering with the sash, the fermenting material usually generates a rather violent heat, which should be permitted to decline to about 90° F., before planting seeds or cuttings in the hotbed. The same protection against excessive heat or cold is used as for the cold-frame; but the hotbed requires much more care in ventilation, since the heating material generates vapor and carbonic acid as well as heat, and these when present in excess are detrimental to plant growth.

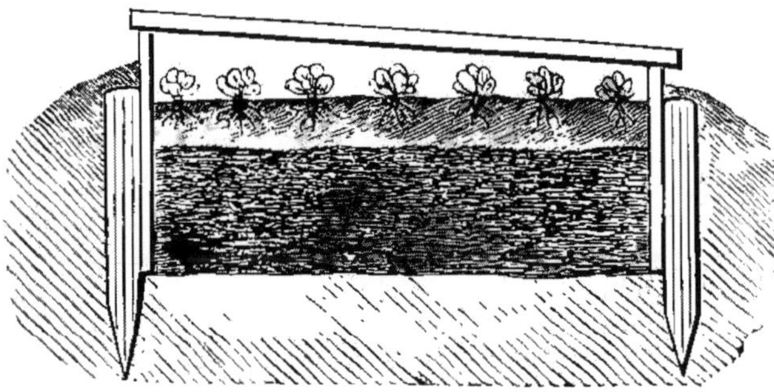

FIG. 94. Cross-section of hotbed in pit. The frame is banked up a little with earth. (After Greiner).

366. The Greenhouse is an expansion of the hotbed, i. e., a structure sufficiently large so that it may be entered, and arranged for heating by fire.* In temperate climates, greenhouses are usually constructed 12 to 22 feet wide, with a gable or M roof, having a slope of 35° to 40°, covered with glass and with the ridge or ridges extending north and south (Fig. 95); but in very cold climates a shed roof facing the south is preferable. Greenhouses are often built with one slope of the roof longer and less steep than the other, and with the ridge extending east and west. Such a roof is called a "two-thirds" or "three-quarters span," according as the longer slope covers two-thirds or three-quarters of the width of the house. The long slope usually faces the south, but houses have recently been built with the shorter and steeper slope facing the south, a plan thought to possess advantages for growing certain plants, as carnations.

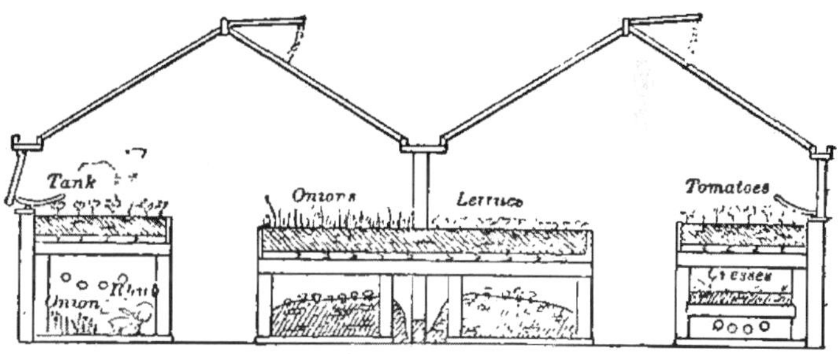

FIG. 95. Cross-section of greenhouse. (After Greiner).

Provision is made for ventilation in glass houses by placing a certain number of movable sash in the roof or elsewhere. In order that the glass may not be far above the plants, the side walls should not exceed five feet in height (241). These may be of wood, but a wall of brick, ten inches thick, with a two-inch air space in the center, is preferable, since this better economizes heat. The furnace and potting rooms obstruct the light least, and

* Hotbeds are now being heated by fire to some extent.

afford the most protection, when located to form the north wall of the house. In houses extending north and south, the south end is usually glazed above the height of the side walls.

367. Heating Devices for the **Greenhouse** are of various kinds. The "smoke flue" is simplest and cheapest in first cost. It consists of a flue extending from the furnace, which is placed somewhat below the floor level, lengthwise through the house, preferably rising gradually to a chimney at the opposite end; or the flue may cross the farther end of the house and return at the other side, to a chimney built directly upon the furnace. The latter method usually gives better draft, since the warmth from the furnace stimulates an upward current of air through the chimney. The flue should be of brick for the first 25 feet from the furnace, as a safeguard from fire. After this it may be of cement, or of vitrified drain-pipe.

Greenhouses of the better class are now almost invariably heated with steam or hot water, or with a combination of the two. Pipes from a boiler located beneath the floor level, extend nearly horizontally about the house, below the benches, returning to the boiler; or the main feed pipe extends overhead to the farther end of the house, where it connects with a system of return pipes beneath the benches. While the steam- or hot-water heating costs much more at the outset than the smoke-flue system,* it is generally found not less economical and far more satisfactory in the long run. Where

* In round numbers, the cost of the smoke-flue may be estimated at ten per cent of the whole outlay required in a house heated by this method, while in one heated with hot water or steam, the cost of the heating apparatus is not far from fifty per cent of the whole.

the pipes need to make many turns, steam is usually more satisfactory than hot water.

368, The Propagating Bed. A certain part of the greenhouse is usually set apart for propagating plants from cuttings. The propagating bed is made upon the ordinary greenhouse beuch, directly over the flue or heating pipes. To furnish the bottom heat (362 a), the space beneath the bench is boxed in with boards. Horizontal doors are, however, provided which may be opened when it is desirable to allow a part of the heat to pass directly into the house. The floor of the bench should not be so tight as to hinder drainage.

In large commercial establishments, entire glass houses are often devoted solely to propagation. Such houses are usually eleven or twelve feet wide, with low side walls. Sometimes lean-to houses are built for propagation, on the north side of a wall, where direct sunlight is cut off.

In making the propagating bed, a thin layer of sphagnum moss is usually spread over the floor of the bench and covered to a depth of two to four inches with well-packed, clean, rather coarse sand, brickdust or powdered charcoal. Sometimes the whole bed is made of moss. These materials are used because they will not retain an excess of water if the proper provision is made for drainage. Sand is most used because it is as a rule readily obtained, but it needs to be selected with care, as it often contains injurious mineral matters. Sand found along the borders of fresh-water streams or lakes may generally be used without washing, but that dug from sandpits should in most cases be exposed to the air for a few weeks, and then be thoroughly washed before being employed for cuttings. The same sand should be used for but one lot of cuttings, as a rule, for it is liable to become infested with fungi that may work havoc with cuttings placed in it.

369. Methods of Controlling Humidity. Where moisture needs to be controlled with especial care, as in propagating delicate plants from green cuttings, or in herbaceous grafting (393), the planted cuttings or the grafted plants are often covered with bell-jars. To guard against sudden fluctuations in temperature, a larger bell-jar is sometimes placed over a smaller one. By means of a bell-jar with a tight-fitting ground plate, evaporation may be wholly prevented from cuttings or plants, if desired. Propagating beds are often covered with glazed sash, in addition to the glass roof of the house, to assist in maintaining a moist atmosphere about the cuttings (Fig. 96).

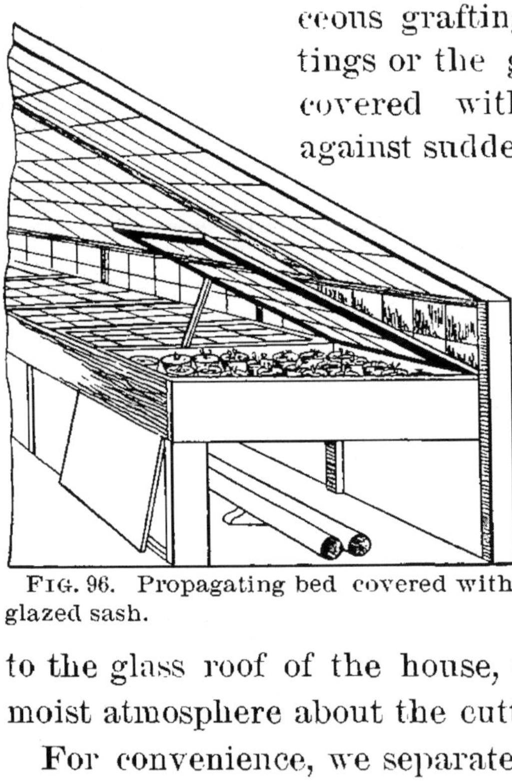

FIG. 96. Propagating bed covered with glazed sash.

For convenience, we separate propagation by cuttings into two divisions, viz., propagation by cuttings from *dormant* and from *active* plants. The requirements of these two classes differ in some respects.

a — *Propagation by cuttings from dormant plants.*

370. The Time to Make the Cuttings. We have seen that plant processes may not be wholly suspended during the dormant period (177). This is true not only of the plant as a whole, but also of detached parts of the plant, if they are protected from evaporation. If cuttings are taken from a plant in autumn and stored during winter in a moist place of moderate temperature, the cut surfaces will partially callus over (73), and the formation of roots or buds may commence before spring.

When new growing points must be developed before the cutting can form a plant, as with cuttings of the stem and roots of many species, cuttings of dormant plants are preferably made at the beginning of the dormant period, i. e., in autumn, and placed during winter under conditions favoring the formation of new growing points.

371. The Storage of Cuttings. Cuttings should be stored in a place sufficiently moist to prevent loss of water by evaporation, and warm enough to favor moderate root growth. Cuttings with ready-formed buds must be kept cool enough to prevent growth of these. Root growth may proceed to some extent at temperatures too low to excite the buds. These conditions are usually fulfilled by covering the cuttings in damp sawdust, sand or loose loam, and storing them through the winter in a moist, moderately cool cellar, or by burying them in the open ground beneath the frost line. In mild climates the latter plan is often preferable. Stem cuttings (373) of plants that do not root freely from the stem are frequently buried with the proximal end (116) uppermost. This gives them, to some extent, the advantage of bottom heat (362 a), since the surface layers of the soil are first warmed by the sun in spring.

Cuttings stored in the ground over winter should be taken up and planted in spring *before the buds expand.*

Cuttings of evergreen plants should not be buried, as this would destroy the leaves, without which they rarely form roots. Cuttings of these plants are usually made in autumn and planted at once in boxes of sand, which are kept for a time in a light, cool place, as a cool greenhouse, until the growing points of the roots have formed, after which they are removed to a warmer location.

372. Planting Cuttings in **Autumn.** Stem cuttings of the currant and other hardy plants, and root cuttings (376) of the blackberry, are sometimes made as soon as the wood is mature in autumn, and planted at once in well-drained loamy or sandy soil in the open ground. Cuttings thus treated often commence to form roots before winter. They should be covered with a little earth and mulched with some coarse litter on the approach of freezing weather, and should be shaded for a time after the opening of spring (Fig. 64).

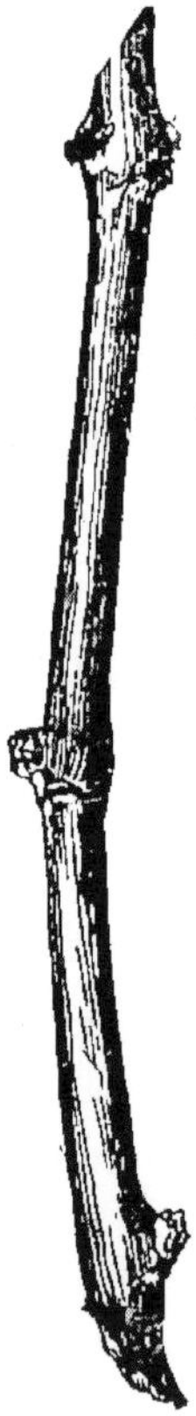
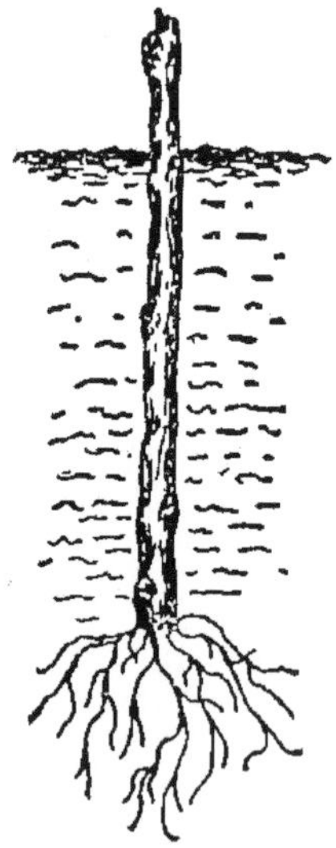

FIG. 97. FIG. 98. FIG. 99.

Fig. 97. Stem cutting of currant.
Fig. 98. Stem cutting of grape. (Both after Bailey).
Fig. 99. Currant cutting rooted.

373. Cuttings from **Dormant Stems** (stem cuttings) usually form roots more promptly if the proximal end is cut off shortly below a node (116). (See Figs. 97, 98 and 99.) In certain plants, as many of the conifers, cuttings root more promptly when cut with a *heel*, i. e., with a small portion of the wood of the previous year at the base. The very short internodes at the junction of the two seasons' growth appear to favor the emission of roots. Some varieties of the grape root more readily when a short section of the parent branch is removed with the cutting, forming a mallet- or T-shaped cutting (*mallet cuttings*).

The cut forming the distal end of the cutting (116) is preferably made somewhat above a node, in order that the bud may not lose an undue amount of moisture by evaporation from the adjacent cut surface.

Cuttings of certain plants that do not readily form roots when made in the ordinary way, may be induced to do so by "ringing" the branch from which the cutting is to be made (428 d), just below a node at about midsummer. Callus will then form at the upper edge of the ring (80), and food will be stored in the stem immediately above it. In autumn the branch may be severed just below the ring and a cutting made, of which the base shall include the callused part, and which may be treated in the usual manner.

374. The Proper Length for **Stem Cuttings** depends upon the conditions under which they are to be grown. Cuttings containing only one bud often root freely and form vigorous plants in the propagating bed, where heat and moisture may be readily controlled. Such short cuttings, however, are seldom used except when cutting wood is scarce. Cuttings intended for planting in the open ground are preferably made at least six inches long.

375. How to **Plant Stem Cuttings.** The general rules given for the planting of seeds apply with nearly equal force to cuttings of the stem (344). Single-bud cuttings should be planted with the bud facing upward, and one-half to three-fourths inch deep, in order that the developing bud may readily reach the surface. Cuttings of more than one bud may be placed upright or at an angle, at such a depth that the bud at the distal end (116) is about on a level with the surface. In cuttings of shrubby plants desired to produce a single stem, the central buds should be rubbed off before planting, leaving but one or two buds at the distal end (Fig. 97).

376. Propagation from **Cuttings** of the **Root.** Plants that naturally sucker from the root (347) and some others may be propagated from short pieces of the root (*root cuttings*). For this purpose roots of about the thickness of a lead-pencil are commonly cut into pieces one to three inches long (Fig. 100), as soon as growth ceases in autumn, and packed in boxes with alternate layers of moist sand or moss. The boxes are preferably stored in a cool cellar where they may be examined from time to time during winter; the sand or moss should be moistened when it appears dry. Root cuttings of different varieties of the same plant often require different degrees of temperature to induce the formation of callus and buds, hence the boxes should be frequently examined, particularly toward spring, in order that those in which the cuttings are backward in starting may be placed in a higher temperature. Thus treated, root-cuttings of many hardy plants, as the plum, raspberry, blackberry, juneberry etc., often form both buds and rootlets by spring, so that they may be planted directly in the open ground. Those of more tender species,

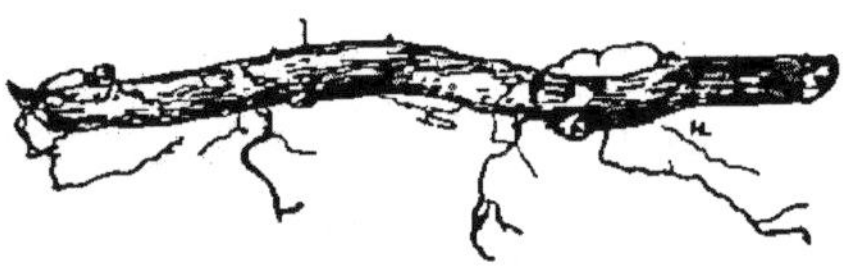

FIG. 100. Root cutting of blackberry. (After Bailey).

as the bouvardia, geranium etc., will not start to the same degree, unless placed in the propagating bed toward spring and given bottom heat.

Root cuttings should be planted shallow, usually not more than one-half to three-fourths inch deep, in order that the developing bud may soon reach the light; otherwise, as in too-deeply-planted seeds, the reserve food may be exhausted before the shoot reaches the surface. When planted in the open ground (372), the soil should be made very fine and carefully pressed about the cuttings; if the weather is warm and dry, shading (Fig. 64) and watering will be necessary.

b — *Propagation by cuttings from active plants* (green cuttings, slips).

377. Nearly All Plants may be **Propagated** from **Green Cuttings.** A succulent cutting of nasturtium * with its leaves intact, and with its proximal end immersed in fresh well- or spring-water, will for a time absorb sufficient of the liquid to make good the loss from transpiration (75). So long as the water remains fresh and the tissues of the stem are unobstructed, the water thus absorbed will answer the same purpose to this cutting as if it had been absorbed by the roots. Food formation (59) will continue, and the growth current (80) will transport the prepared food from the leaves into the stem and in the direction of the roots. No roots being present, however, the growing points of roots will form at the base of the stem, and we shall soon have a rooted cutting. Not all plants, however, root freely in water, possibly owing to an insufficient supply of oxygen therein.

* *Tropæolum.*

With very few exceptions, of which the greenhouse smilax* is one, cuttings of the succulent growth of the stem, with a certain amount of healthy leaf surface intact, will develop roots in all plants, under proper conditions of humidity and temperature; hence propagation from green cuttings is a very common and expeditious method of multiplying plants. The healthy leaf surface, capable of preparing food, is a very important part of a green cutting, because the stem is less abundantly supplied with reserve food during the growth period than during the dormant period (185).

Since the presence of leaf surface upon the cutting greatly promotes transpiration (75), propagation from green cuttings is scarcely practicable in the open air. Bottom heat (362), with a comparatively low air temperature, is especially important with green cuttings, in order that the food prepared in the leaves may be devoted to the formation of roots. A small leaf surface on the cutting is generally preferable to a larger one; in many plants, a portion of a single leaf is sufficient. The leaf surface should in no case be permitted to wilt, hence the cuttings should generally be sprinkled with water as soon as made.

378. Especial Care is **Necessary** in **Propagating** plants from **Green Cuttings.** In planting the cuttings, the material of the propagating bed should be put in close contact with the stems, and no leaves of the cuttings should be covered. Since roots cannot form without oxygen, the bed must not be so freely watered as to exclude all air. Transpiration should be reduced by sheltering the cut-

* *Asparagus meteloides.*

tings from the direct rays of the sun. Movable screens used during sunshine only, are preferable to whitening the glass, which causes too much shade when the sun is not shining.

Damping off, a much-dreaded disease causing cuttings to rot at the surface of the bed, is promoted by excessive heat, over-watering, or insufficient light or air; also by decomposing organic matter in the material of the bed. Affected cuttings should be promptly removed and the trouble corrected.

FIG. 101.

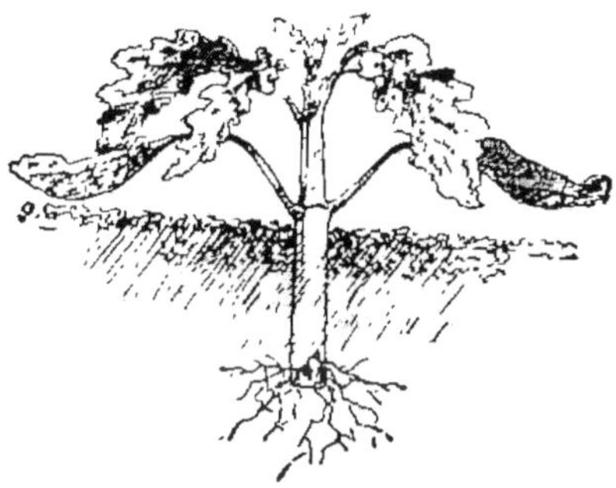

FIG. 102.

Fig. 101. Cutting of chrysanthemum.
Fig. 102. Rooted cutting of coleus. (Both after Bailey).

379. Green Cuttings should be **Potted** as **Soon** as **Roots Form,** which may be detected by their foliage assuming a bright color. They should first be placed in small pots, and until they have commenced growth in these, should be treated precisely as before they were potted.

Propagation by green cuttings includes three divisions, of which the requirements differ in some respects, viz., propagation by cuttings of *herbaceous plants*, of *woody plants* and of the *leaf* or parts of the leaf (*leaf cuttings*).

380. How to **Make Green Cuttings** of **Herbaceous Plants.** In herbaceous plants roots develop most readily from the younger and more succulent parts of the stem. Bend the shoot near its terminus in the form of a **U**, and then press the parts together. If the stems breaks with a snap, it is in the proper condition to root promptly; if it bends without breaking it has become too hard. Cutting below a node (116) is not essential to the formation of roots in herbaceous plants.*

While the propagating house or hotbed is necessary to the extensive multiplication of herbaceous plants by green cuttings, the amateur may readily propagate a limited number of plants by the so-called "saucer system." The cuttings may be placed in glazed saucers containing sand that should be kept saturated with water. The saucers may be set in any warm, well-lighted place, as the window of a living room. The stems being in this case in contact with the water in the bottom of the saucer, the cuttings require less shading than those in the propagating bed.

381. How to Make Green Cuttings of Woody Plants. Cuttings of woody plants are preferably made of harder growths than those best suited to herbaceous plants. They should be selected from young shoots of medium size and from half-mature wood, and should generally contain from two to three nodes, though where the material for cuttings is scarce, single buds may be used in many plants. The base of the cutting is preferably cut shortly below a node, but this is not essential in all plants.

In this kind of propagation a mild bottom heat is helpful, though it is sometimes carried on during the summer months without artificial heat.

382. Propagation by Leaf Cuttings. A considerable number of plants, including the bryophyllum, begonia, gesnera and others, readily develop growing points of

* In a few plants, as the dahlia, the presence of a dormant bud at the crown is essential to the development of the stem the succeeding year. Cuttings of such plants should therefore be made below a node, if the roots are desired for future use.

the stem and roots upon their leaves, a fact often turned to account in propagating these plants. Well-matured leaves, with the principal nerves cut across on the under side, are held in close contact with the surface of the propagating bed by pegging or by light weights, or the leaf may be cut into pieces, which may be placed in the propagating bed and treated as ordinary green cuttings (378).

The leaves of the bryophyllum form rootlets and buds from the notches on their borders wherever these chance to come in contact with a moist medium.

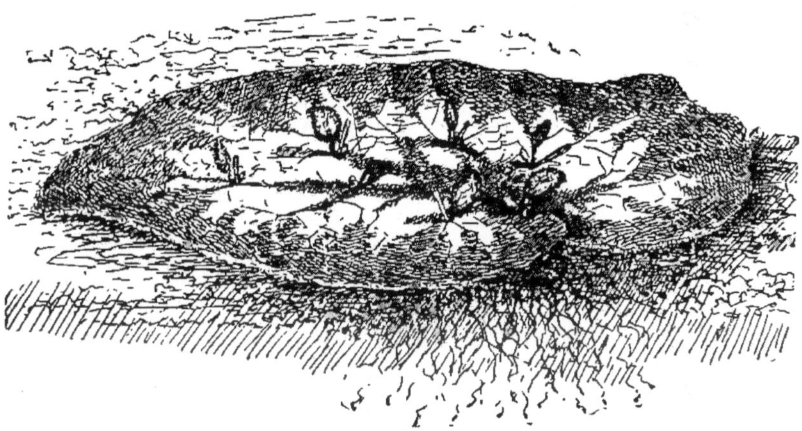

FIG. 103. Leaf of begonia on surface of propagating bed, forming young plants. (After Bailey).

b — *Propagation by Grafting.*

383. Grafting consists in placing together two portions of a plant or of different plants, containing living cambium (69) in such a way that their cambium parts are maintained in intimate contact. If the operation is successful, growth will unite the two parts (70), and plant processes will go on much as if the parts had never been separated. The union usually takes place most rapidly when the cambium cells are in the state of most rapid division, i. e., when growth is most vigorous.

The more intimate the contact of the cambium in the parts brought together, and the less injury their cells sustain in adjusting them, the more likely are they to unite.

The plant that it is desired to change by grafting is called the *stock*, and the part designed to be united to the stock is called the *graft*, *cion* (scion) or *bud*.

Although the tissues of two plants of differing character often unite in grafting, each of the united parts almost always retains its individual character. For example, if one or more buds of the Ben Davis apple are caused to unite by grafting with the stem of a Baldwin apple, the parts that grow thereafter from the Ben Davis buds, though nourished by sap that has passed through the Baldwin roots and stem, with rare exceptions, continue to be Ben Davis, while the parts that grow from the Baldwin stock continue to be Baldwin. To this fact is due the chief value of grafting, viz., it enables us to change the character of a plant.

384. Objects of Grafting. Grafting enables us

a — To change a plant of an undesirable variety into one or more desirable ones;

b — To preserve and multiply plants of varieties that cannot be preserved or multiplied by growing them from their seeds;

c — To hasten the flowering or fruiting of seedlings grown with a view to improving varieties;

d — To change the size of trees, as to make them more dwarf;

e — To restore lost or defective branches;

f — To adapt varieties to special soils;

g — To save girdled trees;

h — To avoid insect injury to the trunk or root, as in grafting the peach on the plum, or the European grape on the American.

385. The Plants that **Unite** by **Grafting.** Plants of different varieties of the same species (21) *almost always* unite by grafting. Examples, the Ben Davis and Baldwin apples, the Bartlett and Seckel pears.

Plants of different species of the same genus (21) *often* unite by grafting. Examples, the peach unites with the plum, many pears unite with the quince; the tomato unites with the potato.

Plants of different genera in the same family or order (21) *sometimes* unite by grafting. Examples, the chestnut unites with the oak; the pear unites with the thorn.

The apparent resemblance of two plants of different species is not always evidence that they will unite by grafting, e. g., the peach and apricot, though resembling each other in many respects, do not readily unite by grafting, but both unite freely when worked upon the plum, though the latter apparently differs from both the peach and apricot more than these differ from each other.

Many plants unite freely when grafted in one direction, that fail to unite when worked in the opposite direction; e. g., many cultivated cherries unite freely when worked upon the mahaleb cherry, while the latter fails to unite when worked upon any of the cultivated cherries; many pears unite freely when grafted upon the quince, but the quince does not freely unite when worked upon the pear. The only sure way of determining what species may be united by grafting is by trial.

Three principal kinds of grafting are in use, viz., *cion grafting*, *budding* and *approach grafting*.

386. Cion Grafting is used in grafting on roots (root-grafting) and very often in grafting on the stem, especially on large trees. The *cion* is a portion of the dormant stem, of the variety it is desired to propagate. It should generally be of the preceding season's growth and should always contain one or more healthy leaf-buds* (132). Cions are usually cut in autumn or during mild weather in winter or early spring, and are commonly stored until needed for use, in a cool cellar packed in moist sawdust, moss or leaves. In climates of severe winters, they should always be cut in autumn. Cions should not be kept so moist as to cause swelling of the buds or the formation of callus (73), nor so dry as to cause shriveling.

In cion grafting, the proximal end of the cion (116) is joined to the distal end of the stock in such a way that the cambium layers of the two coincide in at least one place. Cion grafting in the open air is usually most successful when performed just before or during the resumption of active growth in spring, and the cion is thought to unite more readily if in a slightly more dormant condition than the stock, possibly owing to its more ready absorption of water when in this state.

The joints made in cion grafting are generally coated with a thin layer of *grafting-wax* to prevent evaporation and to keep out water. Sometimes the whole exposed part of the cion is waxed.

387. To make Grafting-Wax for cleft-grafting (392), melt together four parts, by weight, of unbleached rosin,

* Flower-buds are occasionally used, but should be avoided except in special cases.

two parts of beeswax and one part of beef tallow; pour into water, and when sufficiently cool, work with the hands until the mass assumes a buff color; make into

FIG. 104. FIG. 105. FIG. 106. FIG. 107. FIG. 108. FIG. 109. FIG. 110.

Fig. 104. Grafting knife. This should be of excellent steel. The curve in the blade is not essential.

Fig. 105. Cion used for whip-, root- or cleft-grafting, one-fourth natural size.

Fig. 106. Seedling root, used in root-grafting, one-fourth natural size.

Fig. 107. Cion shaped ready for insertion, reduced nearly one-half.

Fig. 108. Portion of seedling root, shaped to receive the cion.

Fig. 109. The cion and portion of root, put together.

Fig. 110. The same as Fig. 109, wrapped with grafting paper.

rolls and wrap with parafined (waxed) paper to prevent the rolls from sticking together. Several other formulas are in use.

For whip-grafting (390), where waxed cord, cloth or paper is used, the beeswax may be omitted from the above formula, or one-half more tallow may be added.

388. Grafting Cord is made by soaking balls of common wrapping twine in melted grafting wax.

389. Grafting Paper is made by painting thin manilla paper with melted grafting-wax. For painting, the paper is preferably spread out on a board of the exact size of the sheet; to prevent too rapid cooling of the wax the board should be heated. The wax should be heated hot enough to spread easily, but not so hot that it is absorbed by the paper. Thin muslin or calico is often used instead of paper.

Grafting paper and grafting cloth should be stored in a cool, moist place to preserve their adhesiveness.

Many kinds of cion grafting slightly differing in details have been described, but the more important are *whip-grafting*, *cleft-grafting* and *side-grafting.*

390. In **Whip-Grafting** (splice-grafting, tongue-grafting) the cion and stock, which should be of about the same thickness, are both cut off with a sloping cut, about an inch long, after which a tongue is formed on each by splitting the wood longitudinally a short distance (Figs. 107, 108). The cion is best cut behind a bud, as shown.

In joining, the tongue of the cion is inserted into the split of the stock, so that the cambium line of the cion and stock (69) coincide on one edge, and the two are crowded together with considerable force, after which the joint is wrapped with a narrow strip of grafting paper or grafting cloth (389), or wound with grafting cord (388). Sometimes the joints are simply tied with unwaxed cord.

Whip-grafting is generally used when the stock is little if any thicker than the cion. It is much used by nurserymen in certain localities in grafting the apple and some other fruits upon roots (root-grafting (391)).

Whip-grafting is also considerably used in some climates of severe winters, in *top-grafting* or "top-working" apple trees in the nursery, in order to give certain slightly-tender varieties the benefit of a specially hardy stock. This grafting is performed on two- or three-year-old trees, that have been grown from root grafts. The trunk is cut off at the height it is desired to form the head of the tree, and a cion of the variety to be propagated is inserted; or several cions are inserted in as many branches. The latter method, while more expensive, has the advantage of giving to the top-grafted trees the branch formation of the stock, which is sometimes important.

As growth starts on top-grafted trees, shoots that push out from the stock should be rubbed off to prevent them from robbing the cions of nourishment.

391. Root Grafting is generally performed in winter and in-doors. The stocks are small trees, grown one or two years from seed (seedlings). These are dug in autumn, and stored as recommended for cions (386). When ready for grafting, the roots are washed and trimmed by cutting off the larger branch roots, after which the stem is cut off at the crown, and the distal end of the root (116) is shaped as directed above (390).

It is then cut off two or three inches down, and the remaining root, if sufficiently thick, is shaped for another stock. Three or four stocks are sometimes made from a single root. As a rule, the stocks should not be less than three-sixteenths inch in diameter, nor less than two inches long.

Some nurserymen prefer to make but a single stock from one root ("whole-root" grafts).

Different nurserymen cut the cions for root-grafts from two to six inches long. In climates subject to drought in summer and severe freezing in winter, the longer cions

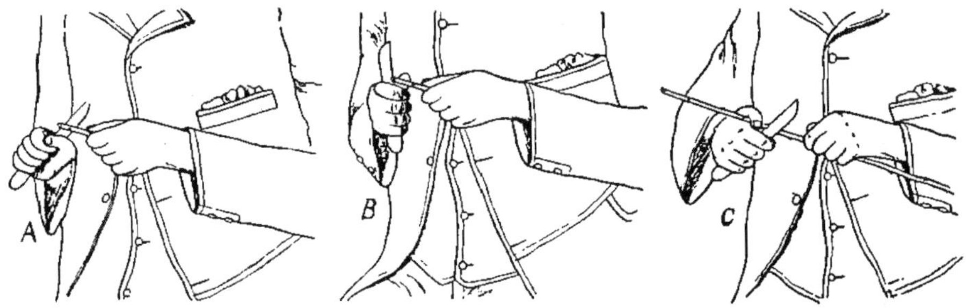

FIG. 111. Shaping the cions for root-grafting. A, making the "long cut"; B, cutting the "tongue"; C, cutting off the cion. These positions, and the movements they indicate, are adapted to rapid work.

are more satisfactory, as they permit the stock to be covered to a greater depth, and encourage rooting from the cion, which is sometimes regarded as an advantage.

Root-grafts should be stored until time for planting out, as directed for cions (386).

392. Cleft-Grafting is generally employed when the stock is considerably thicker than the cion. The cut-off end of the stock is split across its center, with a grafting chisel (Fig. 112), and the proximal end of the cion (116), which is cut wedge-shaped and a little thicker on one edge than the other, is so inserted into the cleft that the cambium of the thicker edge of the cion forms a line with the cambium of the stock (Figs. 113, 114, 115). Success is promoted if the wedge-shaped portion of the cion contains a bud on its thicker edge. When the stock exceeds an inch in thickness, two cions are usually inserted (Fig. 114), to increase the chances of success. The elasticity of the stock should exert sufficient pressure to maintain very close contact between it and the cion; otherwise it should be tightly bound with cord or raffia (393). The cions should contain at least one bud beyond the end of the stock. The

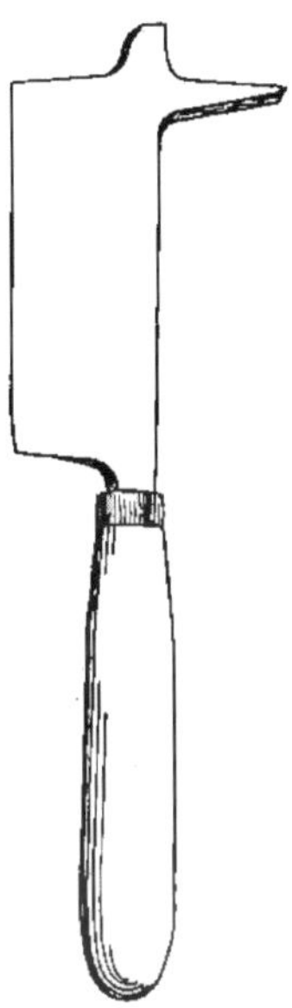

FIG. 112. Grafting chisel for making the cleft in cleft-grafting. The point at the right is for holding the cleft open during insertion of cions. The projection above is for driving this point in or out; one-fifth natural size.

FIG. 113.

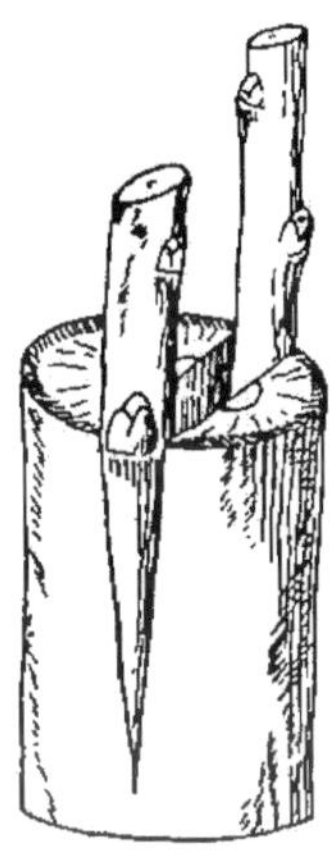

FIG. 114.

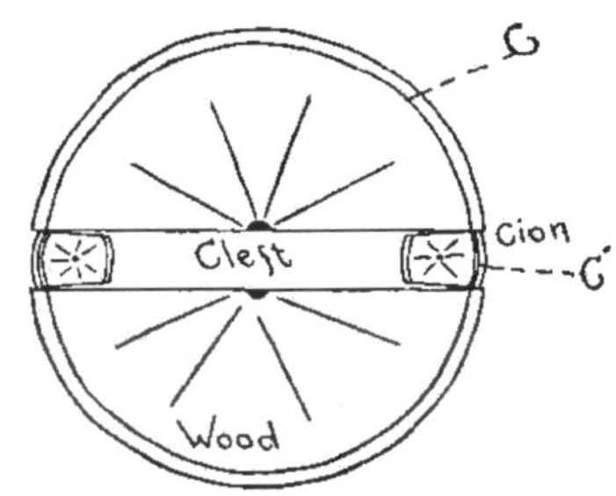

FIG. 115.

Fig. 113. Cion shaped ready for insertion in cleft (After Bailey).

Fig. 114. Cions inserted in cleft, ready for waxing.

Fig. 115. Cross-section of Fig. 113 (After Maynard). C, cambium layer of stock; C′, cambium layer of cion. The cambium layers of the outer edge of the cion should form a continuous line with that of the stock. The cion is made a little thinner at its inner edge to permit the pressure of the stock to be exerted at the outer edge.

wedge-shaped cut is usually made about one inch long, and the cion should be inserted into the cleft as far as the length of the wedge, after which all the exposed wounded surfaces, including the distal end of the cion, should be coated with grafting-wax (387).

Cleft-grafting is most used in top-grafting old trees. Four to six of the main branches, located as nearly eqidistant as possible (Fig. 116), are selected for grafting, and it is desirable to graft these rather near to the top of the trunk.

Branches exceeding three inches in diameter should not, as a rule, be grafted. About half of the top of the tree should be cut away just before the grafting, leaving some branches to utilize a part of the sap. The more or less horizontal branches should generally be selected for grafting, and in these, the cleft should be made horizontally, to give the two cions inserted an equal

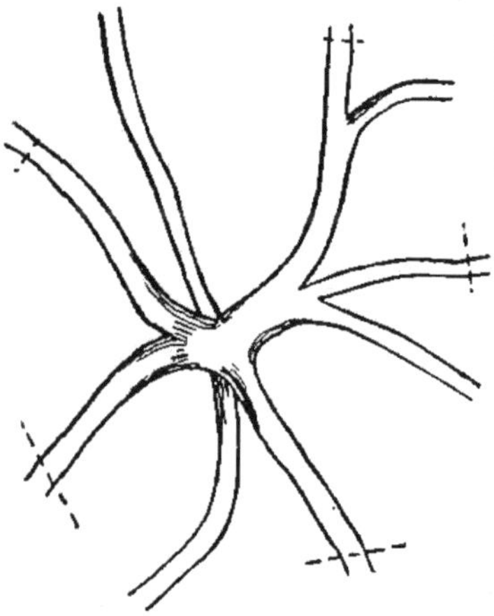

FIG. 116. Branches of tree to be top-grafted, as seen from above, showing where to insert the cions to make a well-formed head, i. e., at the dotted lines.

FIG. 117. Cleft-graft in trunk of old grape vine. The cions are usually inserted below the surface of the ground in grafting the grape, and no wax is used (After Bailey).

opportunity for growth. If both the cions in a branch grow, the weaker one should be pruned off later. As growth starts, shoots from the stock must be rubbed off (390).

The spring following the top-grafting, all or a part of the branches left on the stock at grafting should be pruned off to encourage growth of the grafts. If the tree is large and of a vigorous variety, it is wise to leave a part of these branches until the second spring.

393. Side-grafting is chiefly practiced with plants in leaf, under glass. The cion is joined *at the side* of the stock, which is usually not cut off, and is secured in place by wrapping tightly with bast* or raffia. Three slightly different methods are in use.

a — A shaving of bark, thick enough to reach into the cambium layer, is removed from the side of the stock by making a long vertical cut and a short transverse cut at the base, and to this cut surface the cion is carefully fitted, and bound with raffia. This method is called *veneer-grafting.*

b — A sloping cut is made rather deeply into the sapwood of the stock, into which the cion, after being tapered at its base to the form of a wedge, is inserted (Fig. 118), and the parts are then held closely together by binding with raffia. This method is generally employed in herbaceous grafting, as with the potato, tomato etc. It is also much used in grafting evergreens under glass, and occasionally in grafting outdoor nursery trees. In the latter case, a coating of grafting wax is usually substituted for the tying.

FIG. 118. Side-graft inserted, ready for tying.

c — A short, transverse incision is made, and immediately below this, a somewhat longer, vertical cut — the two cuts, which are just deep enough to reach through the bark, forming a **T** (Fig. 121). The cion is then cut off with a long, sloping cut, and the point inserted, the cut surface inward, beneath the two lips of

* *Bast* is the fibrous inner bark of the bass-wood or linden tree (*Tilia*). It was formerly much used for tying grafts and buds, but has been largely supplanted by *raffia*, which comes from a palm of the genus *Raphia*. Raffia may be purchased of dealers in nursery supplies.

bark formed by the **T**-cut, after which the cion is crowded downward until its cut surface is in contact with the cambium layer of the stock, when the juncture is bound with raffia.

394. Budding is now extensively employed in propagating fruit trees, roses and the varieties of deciduous ornamental trees and shrubs. A (usually dormant) leaf-bud, with a small portion of surrounding bark (Fig. 120), is placed in contact with the cambium layer of the stock. Budding may be successful whenever the cells of the cambium layer are in a state of active division, as indicated by the ready separation of the bark from the wood. In climates having severe winters, budding is most satisfactory when performed near the end of the growing season and with fully-matured buds, in order that the buds may not expand until the following spring; thus the shoots growing from the inserted bud will have the whole season for growth and maturity.

FIG. 119.

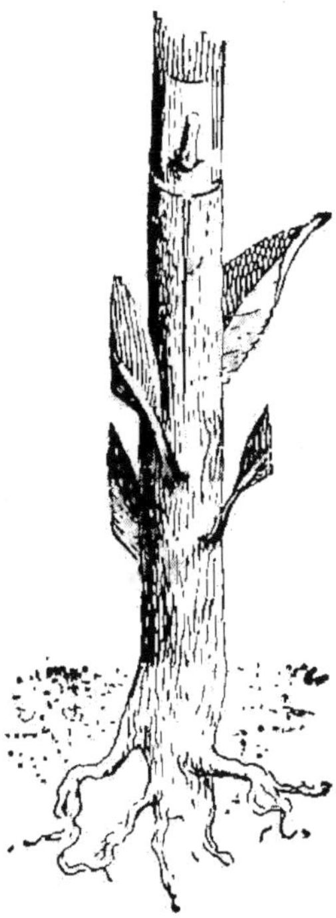

FIG. 121.

FIG. 122.

FIG. 120.

Fig. 119. Shoot containing buds. The white spaces about the buds indicate the amount of bark to be cut off with the bud. The shoot is inverted for cutting the buds.

Fig. 120. Bud cut off, ready for insertion.

Fig. 121. Bud partially inserted between the lips of the stock.

Fig. 122. Bud inserted and tied. (All after Bailey).

With plants that unite freely and with the stock in the proper condition,

395. Success in **Budding Depends Upon**

a — *A fresh condition of the buds*; these must not be in the least shriveled from dryness.

b — *The proper removal and insertion of the bud;* the growing point of the latter (67) must not be injured. If this comes out, leaving the bud-scales partially hollow, the bud will not grow, even if properly inserted. The bud should be inserted promptly to avoid loss of moisture.

c — *The proper wrapping of the wounded bark,* to prevent evaporation and exclude moisture. The ligature should not cover the bud.

d — *The removal of the ligature* after the union, to permit expansion of the stock.

e — *The cutting off of the stock* just beyond the bud, when the latter commences growth, to stimulate its development.

Two methods of budding are in use, viz., **T**- or *shield-budding* and *ring-* or *annular-budding.*

396. In T-Budding, which is the more common and expeditious method, a short shaving, containing a hard and plump bud, cut deep enough to reach through the cambium (Fig. 120), is inserted beneath the bark of the cion, as described for side-grafting (393 c).

The buds, which should be plump and mature and of the variety it is desired to propagate, are taken from shoots of the current season's growth. These shoots ("bud sticks") (Fig. 119) should be cut the day the buds are to be inserted, and should be trimmed at once, and rolled in damp cloth, to prevent loss of moisture. The trimming consists in cutting off the leaves, saving a bit of the leaf stem to serve as a handle while inserting the buds. The stocks, whether grown from seeds or from cuttings, are usually of one or two season's growth. The lower branches of the stock are cut off up to three inches or more from the ground, and a smooth place is selected for the bud, usually on the side least exposed to the sun's rays. With the budding knife, a T-shaped cut is made on the stock (393 c) about two inches above the ground.

FIG. 123. A lesson in budding. The left-hand student is cutting a bud; the central one is lifting the lips of the bark with the spatula of his budding knife; the right-hand student is tying the bud.

A bud is then cut from the bud stick, by inserting the blade of the budding knife about a fourth of an inch below the bud, at such an angle that the back of the blade nearly touches the bark of the stock. The blade is passed just behind the bud, touching the wood, but not removing much of it, and then turned a little, running out about a fourth of an inch above the bud (Fig. 120, p. 220). Often the knife does not run out, but the bark is cut off square, a quarter of an inch above the bud, as indicated in Fig. 119.

With the spatula of the budding knife (397), the lips of bark in the angles of the T-cut are loosened from the wood, when the bit of bark bearing the bud is slipped down behind them (Fig. 121), with the bud pointing upward, until the top end of the bit of bark is just below the horizontal cut of the **T**. Some budders do not use the spatula, but raise the lips of bark with the blade of the budding knife. The center of a strip of moistened raffia is then applied to the stock just below the inserted bud; the ends of the strip are crossed on the opposite side of the stock, brought forward and again crossed just above the bud, thus covering the horizontal cut of the **T**. The ends of the raffia are then brought behind the stock, tied in a half knot, and drawn moderately tight (Fig. 122), pressing the lips down snugly about

FIG. 124.

FIG. 125.

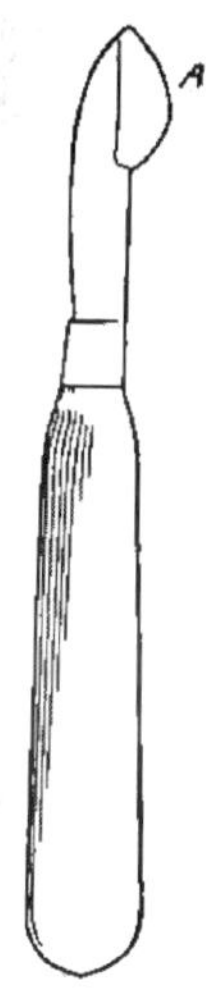

FIG. 126.

Fig. 124. Man budding in nursery row (After Bailey).

Fig. 125. Budding knife with ivory spatula on the end opposite the blade.

Fig. 126. Budding knife made from erasing knife by rounding the edge at A.

the bud, which now protrudes between the lips.

If the bud "takes," it will unite with the stock in a few days. The raffia should be taken off in about ten days, by cutting it on the back side of the stock, to enable the latter to expand by growth.

397. The Budding Knife should contain a blade of good steel, shaped as indicated in Fig. 125, and a round-edged spatula for lifting the bark. The spatula is better placed on the back of the blade, as shown in Fig. 126.

398. Ring Budding is used to some extent in the propagation of thick-barked plants, as the hickory and magnolia. A section of bark is removed nearly or entirely around the stock, and a similar section containing a bud from the variety it is desired to propagate, is fitted to its place and snugly bound with raffia. Ring budding is oftener performed in spring than later in the season.

399. Approach Grafting is now seldom employed, except in a few plants that unite poorly by other methods. It is only possible between two plants in

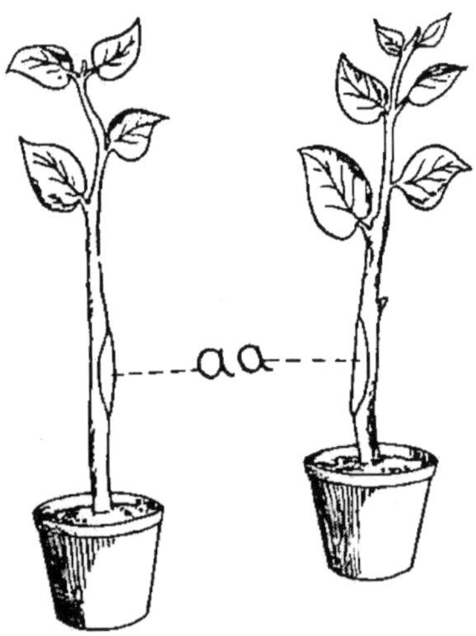

FIG. 127.

FIG. 128.

Fig. 127. Two plants prepared for approach grafting. The cut surfaces a a, are to be placed together and bound.

Fig. 128. Two plants bound together for approach grafting (After Bailey).

close proximity, or between parts of the same plant, since the graft is not severed from the parent until it has united with the stock. The plants are nourished by their own roots until the union takes place.

Approach grafting is performed during or just previous to the growing season. The parts are held in contact by binding them with raffia; the juncture should also be waxed if the work is done in the open air.

Two methods of approach grafting are in use:

a — A shaving reaching into the cambium layer is removed from both stock and graft on the sides toward each other (Fig. 127), and the cut surfaces are brought together and closely bound until they unite (Fig. 128), after which the graft is cut off below, and the stock above, the union.

b — The top of the stock is cut off with a long sloping cut, preferably behind a bud, and the cut surface of the remaining part is inserted beneath the bark of the graft, as described in side-grafting (393 c), except that the **T**-cut is inverted, and the stock is inserted from beneath.

The graft is cut off below the point of union when the parts are fully united.

In both these methods the graft should be severed gradually to avoid a check to the growth.

SECTION II. TRANSPLANTING

400. Transplanting consists in lifting a plant from the medium in which its roots are established, and in replanting the latter in a different location. Transplanting is a violent operation because the younger roots with their root-hairs that absorb the greater part of the water required by the plant (102) are, as a rule, largely sacrificed in the lifting process. The water supply, so vitally important to the plant (63), is thus greatly curtailed until new root-hairs can be formed.

Vigorous plants are generally better able to endure transplanting than feebler ones, because they can sooner repair the damage done to their roots. It follows that plants endure transplanting with less facility as they advance in age beyond the period of greatest vigor (9).

401. The Most Favorable Time for **Transplanting,** in the case of plants that live more than one year, is during the dormant period, because growth processes are then least active, and comparatively little water is needed. In countries having mild winters, the most favorable time for transplanting is generally at the beginning of the dormant period, provided this comes at a moist season of the year. The roots will then have time to slowly callus over their wounds and to form new rootlets, and thus be prepared for active growth in spring. But in countries of severe winters, where the roots are largely frozen in the soil for two or three months, and in countries in which the autumn is generally dry, spring is, as a rule, the more favorable season for transplanting.

Trees that have been long exposed to cold, drying winds and have thus suffered depletion of water from their buds and branches, are better not lifted until the buds begin to swell. This is especially true of evergreen trees in severe climates. Being always in leaf these require more careful treatment than deciduous trees.

We shall consider transplanting under three divisions, viz., a, *lifting the plant;* b, *removing the plant;* and c, *replanting the plant.*

A — Lifting the Plant.

402. The object to be attained in this operation should be to remove the roots from the soil with the least possible damage consistent with reasonable economy of time and labor. Plants in low vigor should receive especial care in this respect. Very young plants, as of tobacco, cabbage, lettuce etc., grown thickly in the seed-bed, are often pulled from the soil with the hands. In this case, the soil of the bed should first be saturated with water, in order that the roots may be broken as little as possible, and may come up with more or less adhering soil. It is generally preferable to grow such plants in drills rather than broadcast. This enables them to be drawn from the soil with less damage to their roots.

Trees and shrubs sufficiently grown for their final planting out should be more carefully handled. If it is necessary to cut off the main roots, the farther from the trunk this is done, the better for the tree, and the spade used should be kept as sharp as possible. *The roots should not be barked, mangled or split by the digging tools,* as is so often done with nursery stock. When possible, one person should lift on the tree or shrub, while another removes the earth from about the roots. Tree-digging machines are now much used by the larger nurserymen.

403. Lifting Large Trees. Trees considerably larger than nursery sizes are best lifted when the ground is frozen about their roots. A trench may be dug about the tree before the ground freezes, deep enough to permit the severing of the main roots, and a hole for the reception of the cylinder of earth left within the trench should also be dug at the place to which it is desired to remove the tree. This cylinder should be large enough so that the tree is left with abundant roots, or as large as can be removed with the apparatus at hand. When the ground is frozen to the proper depth, the tree may be tipped over by means of a rope and windlass, after which the cylinder of earth enclosing the roots may be pried up sufficiently to allow some low vehicle to be placed beneath it. The branches are usually permitted to drag upon the ground in removal, as the wounded parts may be cut off in the severe pruning necessary in planting large trees (409c).

Large trees may be lifted or lowered to accommodate grading. A trench is dug around the tree, leaving a cylinder of earth intact about the roots. Soil is then removed from beneath one side of the cylinder below the roots and a block set under as a fulcrum. The top of the tree is then inclined toward the fulcrum by means of a rope, until the roots are lifted on the opposite side. If the tree is to be raised, soil is packed under the elevated roots, after which the top is tilted in the opposite direction, until the roots are lifted on the fulcrum side, when soil is placed under as before. This process is repeated until the tree has been lifted to the desired height. If the tree is to be lowered, earth is removed at each tilt.

404. Sacking the **Earth-Enclosed Roots** is practiced in lifting and removing orange trees in California and may be profitably employed with other evergreens. A rather deep trench is dug at one side of the tree, and from this trench, the deeper roots are severed. The top earth is then removed down to the first lateral roots, when all the remaining large roots are severed at some distance from the trunk. The tree is next tilted to one side and a piece of burlap or matting is drawn beneath it, after which the matting is folded about the earth cylinder and well tied.

B — Removing the Plant

Plants with their roots out of the soil should be carefully protected from mechanical injury, from drying and from freezing. To insure such protection, plants to be transported any considerable distance should be *packed.*

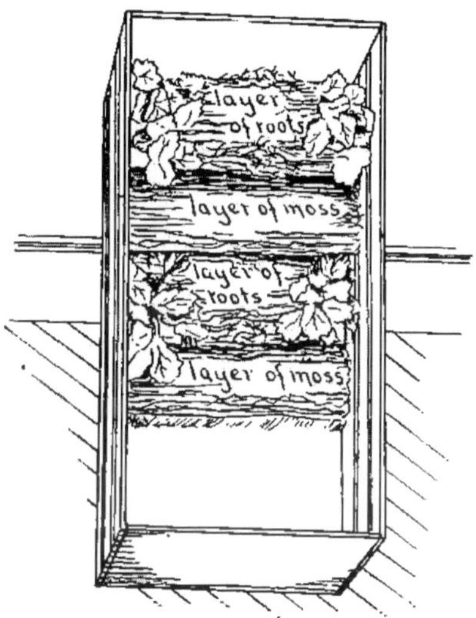

Fig. 129. Showing how plants should be packed for shipping.

405. Plants Packed for **Transportation** should be inclosed throughout, and the roots should be in close contact with some moist material, preferably bog moss. Straw is often used for this purpose and answers well for packing about the trunks and branches of trees, but it is inferior to moss for enclosing roots, as it is more liable to heat and does not so well retain moisture.

Herbaceous plants, as the strawberry, cabbage, sweet potato etc., may be packed in layers separated with moss, as follows: Over the bottom of a box, the width of which is about twice as long as the plants to be packed, and which has slatted sides, place a thin layer of damp (not wet) moss, and over this, place a layer formed of a double row of the plants, with their roots at the center, overlapping a little, and the tops toward the sides of the box (Fig. 129). Then put in another layer of moss and of plants, and so on until the box is full, or the desired quantity is packed. The thickness of the layers will depend upon the time of year, the temperature, the distance to be transported and the kind of plants. The warmer the weather, the thinner should be the layers of plants, as a rule. When the top of the box is put on, the contents should be pressed sufficiently to prevent the plants from shaking out of place.

406. Puddling the **Roots** of **Trees,** i. e., dipping them in a paste of soil and water, is much practiced by nurserymen and tends to prevent them from drying. The paste should be made with rather light, loamy soil and of the consistency of cream.

407. Trees are commonly **Bundled** for **Transportation** to economize space. For this purpose, a device resembling a sawbuck, with the arms cushioned with burlap or carpeting is very convenient. The trees are laid between the arms, with the roots placed evenly at one end. The stems are then drawn snugly together with a broad strap, after which they are bound with soft cord or with young and tender shoots of the osier willow.* After bundling, the space between the roots should be filled with damp moss, and the whole mass of roots surrounded with the same material. If the distance to be transported is short, the mossed roots may be sewed up in burlap or matting and the tops may be tied up in straight straw, or the whole bundle may be enclosed in burlap. If the distance is long, the bundle should be boxed, to more effectually prevent the trees from damage. The bundles may be packed very closely in the box without

* *Salix viminalis.*

injury, provided they nowhere come in direct contact with it. Boxed or bundled trees, that cannot be shipped at once, should be stored in a cool, damp place.

408. Unpacking and **Heeling-In.** Packed plants should generally be removed from their package as soon as they reach their destination. If they cannot be replanted immediately, they should be *heeled-in.* This consists in re-

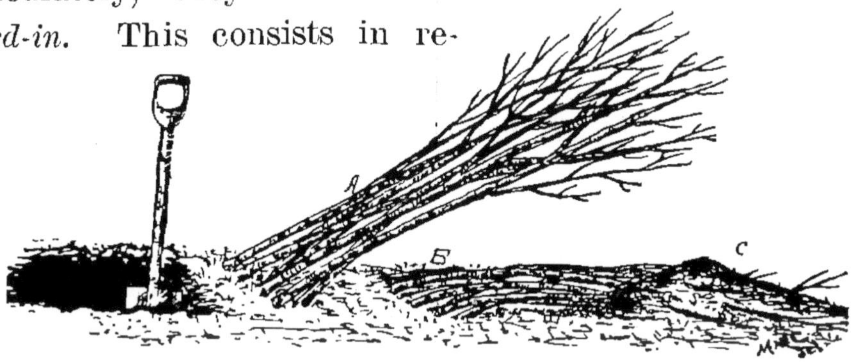

FIG. 130. Nursery trees heeled-in to prevent drying. A, a short row of trees with only the roots covered. B, a row with their tops bent down and covered with earth at C (After Green). Sometimes the whole tops are covered. *Trees should not be heeled-in in the bundles.*

moving them from their bundles and temporarily planting their roots in soil (Fig. 130). The roots should be well covered, and if at a dry season, they should also be mulched. To avoid mixing varieties, a separate row should be made of each sort.

Nursery trees that cannot be packed for shipment at the proper time, are often lifted and heeled in, to retard the starting of the buds.

C — REPLANTING

409. Preparation of the **Plant.** a — *Washing the roots.* The "puddled" roots of nursery trees (406) are sometimes found inclosed at unpacking in a mass of mud that is so compact as to largely exclude the air (Fig. 131).

The roots of such trees should be washed clean before replanting (Fig. 132).

b — *Trimming the roots.* The roots of trees that have been broken or mangled in the lifting or transportation, should be cut back to sound wood with a sharp knife.

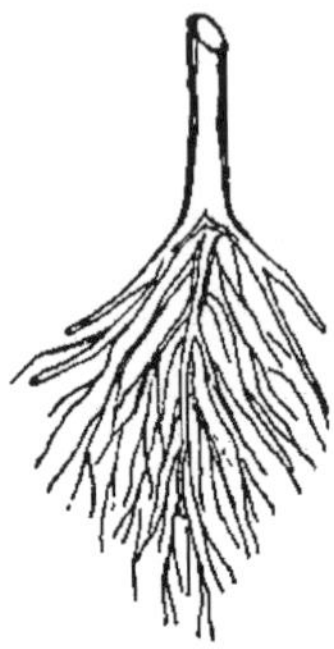

FIG. 131. FIG. 132.

Fig. 131. Puddled roots of nursery tree.

Fig. 132. The same washed, ready for planting.

Fibrous rooted plants, as the strawberry, are much more readily planted when the roots are trimmed, as shown in Fig. 31, (p. 73).

c — *Reducing the top.* The buds of trees and shrubs should generally be reduced in number at replanting to correspond with the destruction of the younger roots during the lifting process; otherwise the water supplied by the roots may be insufficient to open the buds (63). This is best accomplished by thinning out and

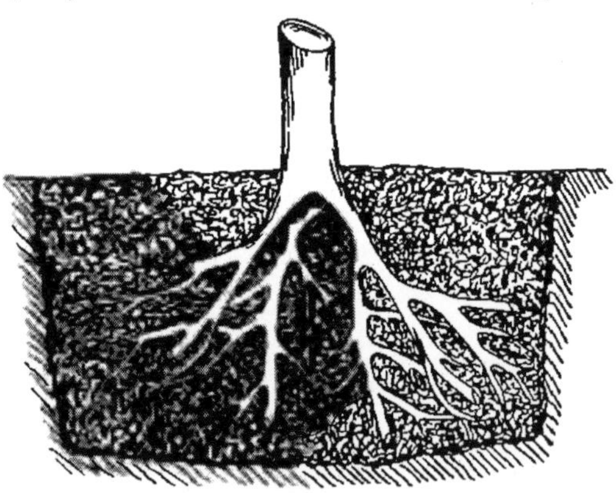

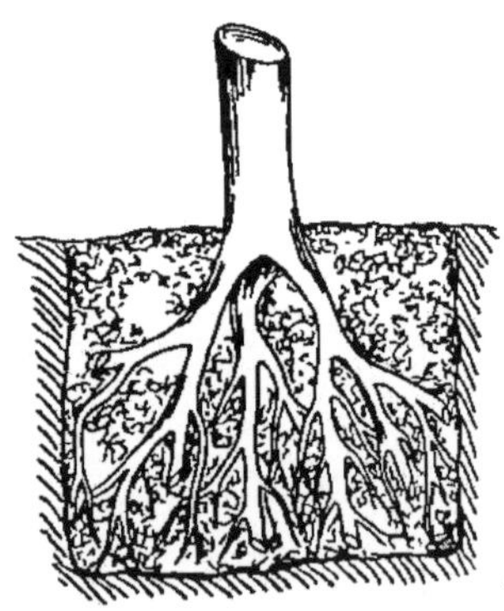

FIG. 133. FIG. 134.

Fig. 133. Roots of tree properly planted.

Fig. 134. Same improperly planted.

cutting back the branches. As a rule, it is better to reduce the top rather sparingly at replanting, with the expectation of cutting it back further if the buds do not promptly opeu at the proper time. The branches that can best be spared should be removed (420). Failure to properly reduce the top is a frequent cause of death or loss of vigor in transplanted trees.

Small plants in leaf, as the strawberry, cabbage etc., usually endure transplant-

FIG. 135.

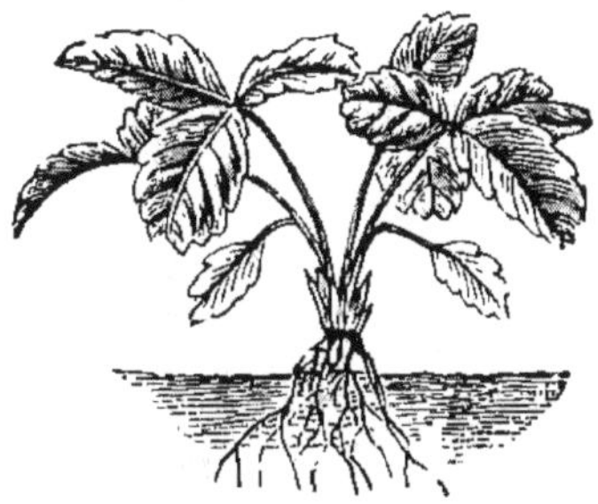

FIG. 136.

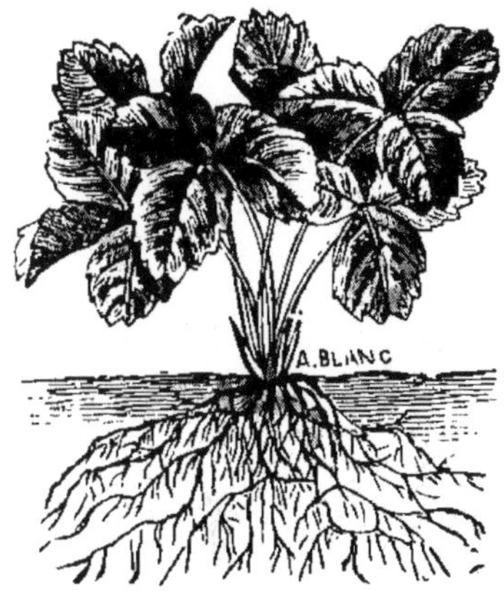

FIG. 137.

Fig. 135. Strawberry plant too deeply planted.
Fig. 136. The same planted too shallow.
Fig. 137. Strawberry plant properly planted.

ing better if their larger leaves are removed at replanting.

d — *Wetting the roots* just before replanting is quite important, as it favors intimate contact with the soil particles.

Plants that have suffered from loss of moisture in transit should have their roots soaked in clean water for a few hours before replanting. Deciduous trees of which the bark is considerably shriveled may often be saved, if the center of the buds is still fresh, by burying them in moist earth until the bark resumes its plumpness.

410. Replanting the **Roots.** The object to be attained in this operation is to *place moist and well-aerated soil in contact with all of the roots* of the plant. The roots should also be placed at about the same depth, and in nearly the same position that they grew before the removal.

Fig. 133 shows the roots of a tree properly planted. The hole was dug sufficiently large so that the roots were readily placed in it without crowding, and the soil was so well worked in among the roots that it comes in contact with their whole surface.

Fig. 134 shows the roots of the same tree improperly planted. The hole was dug so small that the roots were necessarily crowded out of their natural position, and the earth was thrown in so loosely that it comes in contact with only a part of the root surface. Distortion of the roots of trees and shrubs at planting may cause injurious root galls.

In planting trees of which the roots are not already inclosed in soil (403), the hands should be freely used to bring the soil in contact with the whole root surface, and the earth should be moderately packed about the roots with the feet, or otherwise.

If the soil is dry, it is probably better to moisten it before placing it about the roots, rather than after, as we have then a better opportunity to judge of the quantity of water required, and the soil is less likely to settle away from the roots.

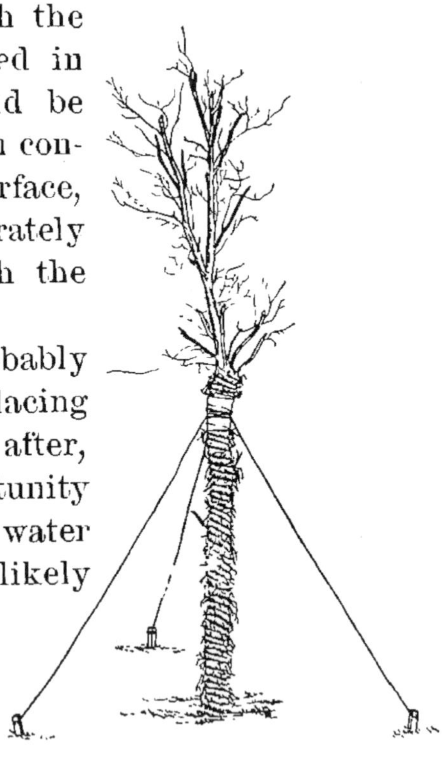

FIG. 138. Large transplanted tree wound with hay rope and supported by wires.

Trees of considerable size should generally be staked or otherwise supported after planting, to prevent shaking by wind (Fig. 138). Surrounding the trunk with poor-conducting material as hay, straw or canvas, tends to prevent damage from sun-scald

(186), to which recently-transplanted trees are especially liable; as the evaporation stream (78) is much reduced, the bark tends to become unduly heated.

411. Devices for **Transplanting.** With young trees and plants, that possess abundant vigor, rapidity of planting is often of greater importance than the observance of precise rules. In this case, that method is best which secures a given number of transplanted and vigorously-growing plants at the least cost. The transplanting devices shown in Figs. 139–141, inclusive, aid greatly in accomplishing this end.

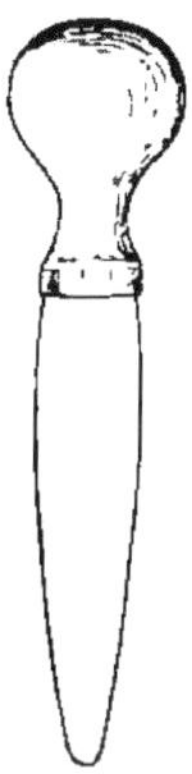

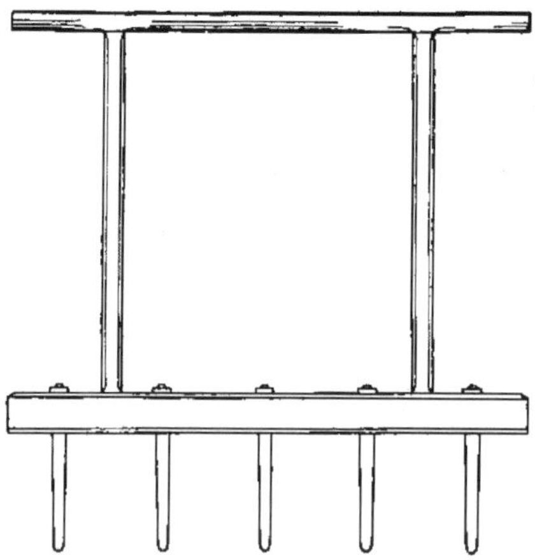

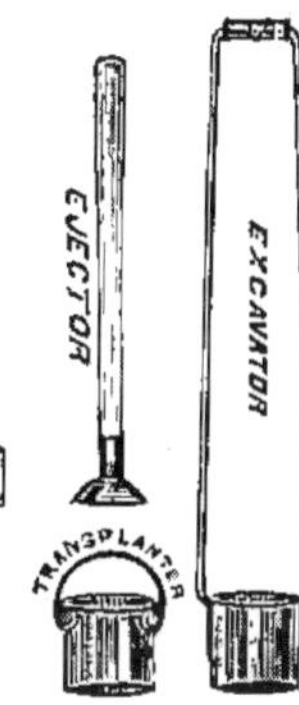

FIG. 139. FIG. 140 FIG. 141.

Fig. 139. Flat steel dibber, (one-sixth natural size).
Fig. 140. Tool for planting root grafts and cuttings (much reduced).
Fig. 141. Richards' transplanting tools, made by F. Richards, Freeport, N. Y.

The *dibber* (Fig. 139) is perhaps, aside from the spade, the most valuable single tool for transplanting. It is used for opening the hole to receive the roots of small plants, as cabbage, celery, onions etc., and for pressing earth about the roots; it answers equally well for planting cuttings and root grafts. The manner of using it appears in Figs. 143 and 144.

Fig. 140 shows a very convenient tool for planting root grafts and cuttings. It consists of six steel dibbers, attached in a line to a piece of scantling, at the distance apart at which the plants or cuttings are to be planted, with a handle affixed above. In using this tool, the operator crowds the dibbers into the soil with the foot, guided by a line. He then moves the frame to and fro until the holes are sufficiently opened, when he withdraws the dibbers by lifting the frame, and passes on to repeat the operation. A person follows inserting the grafts or cuttings, and crowding earth about them with the ordinary dibber. See also Figs. 145 and 146.

Fig. 141 shows a set of transplanting tools, useful in removing a limited number of plants that are not closely crowded and that need to be carried but a short distance. They are especially useful for transplanting strawberry plants during summer and autumn. These tools and also the Baldridge transplanter enable the plant to be readily lifted with a cylinder of earth and replanted in a hole just large enough to receive the latter.

FIG. 142. Bemis Transplanter, made by Fuller & Johnson Manufacturing Co., Madison, Wis.

Fig. 142 shows a successful machine for planting tobacco, cabbage, strawberry and other low, herbaceous plants. It plants these as rapidly as two boys can deliver them to it in the proper position, and waters the soil about the roots at the same time.

412. Potting and **Shifting.** *Potting* is the act of planting plants in greenhouse pots.

The pots should be clean and are usually dipped in water before receiving the plants, until they have absorbed as much of the liquid as they will take without leaving any upon the surface. Rooted cuttings are generally potted in pots one and one-half or two inches in diameter, and the plants are changed to larger pots (*shifted*) as the roots require more room. Pots three inches or more in diameter are commonly filled one-third full or less with pieces of broken pots (potsherds) to

FIG. 143.

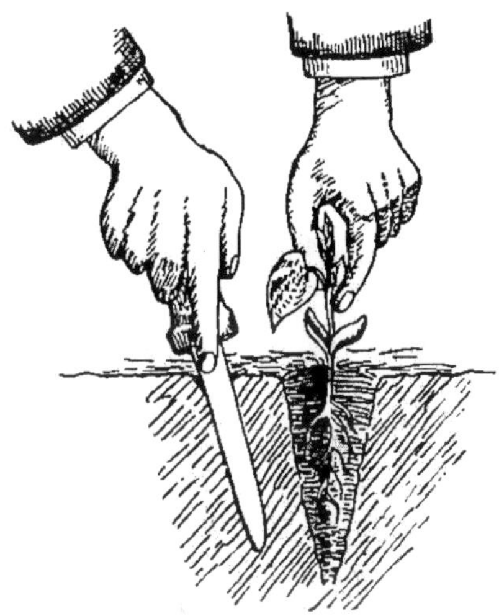

FIG. 144.

Showing manner of using the dibber in planting.

Fig. 143. Inserting roots in the hole opened by dibber.
Fig. 144. Pressing earth about roots with the dibber.

insure abundant drainage, and these are often covered with a little sphagnum moss before putting in the soil. The soil used for potting should be of a sort that does not harden, "bake," on drying, and should generally be liberally supplied with plant food. Decayed sods from an old pasture, leaf mold, decomposed manure, and sand, the whole mixed and sifted, form a good potting soil. The proportions of the different ingredients used vary with different plants. The soil should be moderately moist, and should be closely pressed about the roots. The details of potting are shown in Figs. 147 to 150.

Shifting is the changing of a plant from one pot to another, usually a larger one. Plants in small pots are generally shifted as often as their roots begin to crowd, and the shifting is continued as long as further growth is expected. When bloom is desired, the pots are permitted to become filled with roots (136).

FIG. 145. FIG. 146.

Rapid method of planting strawberry plants with spade.

Fig. 145. One man opens the hole by inserting the spade, back side forward, and crowding it toward him. The other inserts the plant, taking care to spread out the roots well.

Fig. 146. The man withdraws the spade and crowds the earth closely about the roots of the plant with his foot.

The pots into which plants are to be shifted should be prepared as directed for potting. A little potting soil is placed in the bottom of the pot, or over the drainage material, after which the plant to be shifted is tipped out of its pot, by inverting the latter, placing the hand upon the surface of the soil, to support it, and tapping the rim of the pot gently upon the edge of the potting bench.

If the soil is in proper condition, it will readily slip out of the pot intact, after which it should be placed in the center of the new pot and the space about it filled with potting soil moderately pressed down. The roots of woody plants should not be covered deeper than they grew before the shifting. See Figs. 152, 153 and 154.

FIG. 147.

FIG. 148.

Fig. 147. The workman takes a pot in his left hand, and at the same time a handful of potting soil in the right hand.

Fig. 148. He places the soil in the pot, pressing it against one side with the right hand, while he picks up a plant with the left hand.

D — AFTER-CARE OF TRANSPLANTED STOCK

413. Mulching the soil about transplanted plants (233) is very important in localities subject to drought. As a rule, it is wise to apply the mulch immediately after transplanting, but with trees transplanted very early in spring, it is better to defer mulching until the soil becomes sufficiently warm to promote root absorption (102).

Watering recently-transplanted plants requires discretion. As a rule, mulching is preferable to watering, but if mulching proves insufficient, watering is the last resort. In this case, the soil about the roots should be saturated with water and should not be permitted to become dry again until growth starts. A hole may be made in the soil about the roots and kept filled with water until the liquid ceases to soak away rapidly, after which it should be occasionally filled until growth commences.

FIG 149. FIG. 150.

Fig. 149. Placing the roots of the plant against the soil in the pot with the left hand, he takes another handful of soil with the right hand.

Fig. 150. He fills the remaining space in the pot with soil and presses it down with the thumbs, tapping the pot gently upon the bench in the meantime.

414. Shading plants transplanted in leaf, until the roots resume activity, is important (236). Evergreen trees and shrubs may often be shaded with barrels or boxes, or with boughs from other evergreen trees.

415. Tardy Starting into **Growth** after transplanting is usually evidence that the roots are not supplying sufficient water. In such cases, if other precautions have been observed, it is well to further reduce the top. Plants in this condition may sometimes be saved by wrapping the stem in oiled or rubber cloth to check loss of moisture, or with straw or moss which may be wet frequently till growth starts.

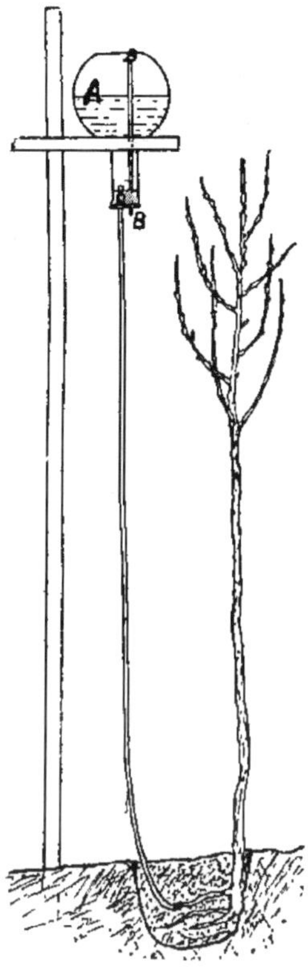

FIG. 151.
Device for starting growth in trees.

The device shown in Fig. 151 often causes recently planted trees to start growth that seem likely to fail without it. It consists of a flask or bottle containing distilled, or rain water, supported a few feet above the ground and connected by a rubber tube with the cut-off end of a root, as shown. If the inverted flask is used, a short tube B B should extend through the cork and to near the bottom of the flask, to admit air.

Flower-buds should generally be removed from recently-transplanted plants (140).

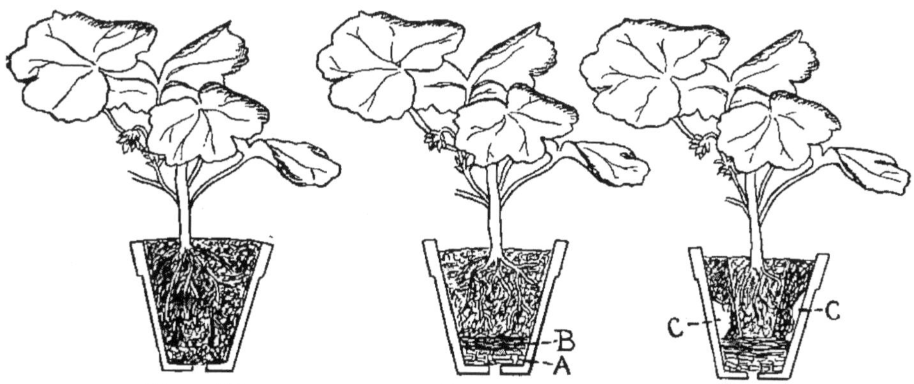

FIG. 152. FIG. 153. FIG. 154.

Fig. 152. *A poorly-potted plant.* No provision is made for drainage; the pot is filled to the top with soil, leaving no space to receive the water; and the stem of the plant is not at the center of the pot.

Fig. 153. *A well-potted plant.* A, potsherds; B, moss.

Fig. 154. *A poorly-shifted plant.* C, open spaces due to insufficient pressing of the soil.

SECTION III. PRUNING

416. Pruning is the removal of a part of a plant, in order that the remainder may better serve our purpose.

The parts of plants, being less highly specialized than those of animals, may be removed with less damage to the individual than is possible with animals, except in the lowest types.

The word *pruning*, as commonly used, applies chiefly to the removal of parts of woody plants with the knife, shears or saw, but the operations defined below properly come under the same head.

a — *Pinching* is the removal with the thumb and finger of the undeveloped nodes at the terminus of growing shoots, in order to check growth.

b — *Trimming* or *dressing*, when applied to young nursery stock, is the shortening of both roots and stem, preparatory to planting in nursery rows. The roots are shortened to facilitate planting, and the stems are shortened to reduce the number of buds (409 c).

c — *Topping* is the removal of the flower stalk, as in tobacco, to prevent exhaustion of the plant by the formation of seed.

d — *De-tasseling* is the removal of the staminate flowers, (tassels) of undesirable plants of Indian corn, to prevent pollination from them (151).

e — *Suckering* is the removal of shoots that start about the base of the stem, or in the axils of the leaves, as in Indian corn or tobacco. Its object is to prevent exhaustion of the plant by the production of needless shoots.

f — *Disbudding* is the removal of dormant buds, to prevent the development of undesirable shoots.

g — *Ringing* is the removal of a narrow belt of bark about a branch, to obstruct the current of prepared food (138).

h — *Notching* is the cutting of a notch just above or below a bud or twig to modify its growth.

i — *Thinning fruit* is the removal of a part of the fruits upon a plant, to permit the remaining ones to attain larger size, and to prevent exhaustion of the plant by excessive seed production.

j — *Deflowering* or *defruiting* is the removal of flower-buds or fruits to prevent exhaustion of the plant (140).

k — *Root pruning* is the shortening of the roots of plants in the soil, to check growth, or to stimulate the formation of branch roots nearer the trunk (105).

417. The Season for **Pruning.** The milder kinds of pruning, as pinching and disbudding, may be performed whenever the necessity for them appears. But in perennial plants, severe pruning, as the removal of branches of considerable size, is generally least injurious if performed during the dormant period. As the exposure of unhealed wounds may cause damage from drying, and invites infection by injurious fungi (321), severe pruning is commonly best performed toward the end of the dormant period, i. e., in early spring because healing is most rapid at the beginning of the growing season (73). Pruning should not, however, be done at a time when sap flows freely from wounds, as this tends to waste reserve food. In plants subject to this, as the maples and grape, pruning is probably best performed just before the sap-flowing period.

418. Where and **How** should the **Cut** be **Made** in **Pruning?** Since the movement of prepared food is from the leaves toward the root (80), it follows that when a branch is cut off at some distance from the member that supports it, the wound cannot heal, unless there are leaves beyond the wound to manufacture food, and thus make a growth current possible (73). The cut should, therefore, be made close enough to the supporting member so that it can be healed from the cambium of the latter. In woody plants, there is usually a more or less distinct swelling about the base of a branch (Fig. 155), produced by the

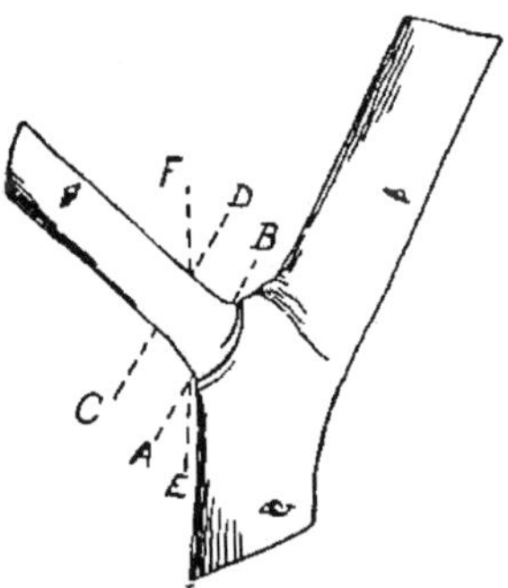

FIG. 155.

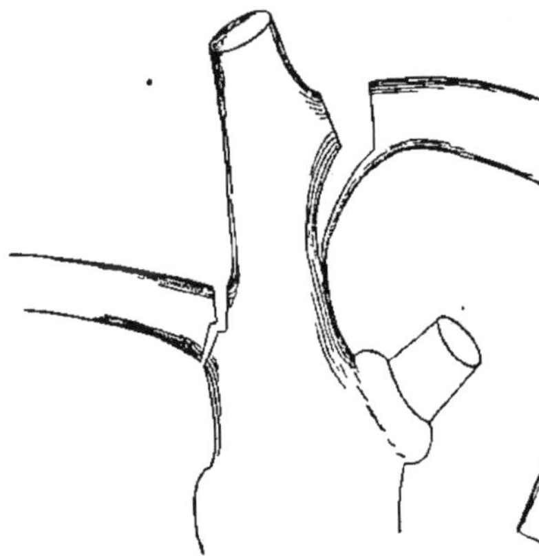

FIG. 156.

FIG. 157.

Fig. 155. Showing *the proper place* to make the cut in pruning. A wound made by a cut on the dotted line A-B will be promptly healed. One made on the line C-D or E-F will not. In Fig. 156 the lower branch was cut off too far from the trunk.

FIG. 156. Showing *how* to make the cut in pruning large branches. The upper cut, all made from above, permits the branch to split down. The left cut, first made partly from below, prevents splitting down.

Fig. 157. Pruning to an *outside* or *inside* bud. Cut as in the figure, the uppermost bud would form a shoot that tends to vertical. Cut on the dotted line, the uppermost bud would form a shoot tending to horizontal.

cambium of the supporting member and just beyond this swelling, a more or less distinct line marks the point where the cambium of the branch and of the supporting member unite. In a healthy tree, a wound made by a branch of reasonable size, cut off at this line, will usually heal promptly, but if the cut is made much further out, it will not.

The cut should generally be made at right angles with the branch, rather than parallel to the supporting member, since it is important that the wound be no larger than is necessary. Wounds so large that they cannot heal promptly should be painted with lead and oil paint to preserve the wood.

419. Unhealed Wounds Introduce Decay into the heartwood of trees. Since the cells of the heartwood contain no protoplasm (72) and are always moist, they form a congenial field for certain destructive fungi (321), that having once gained entrance, sooner or later destroy the heartwood of the whole trunk, thus greatly weakening it and preparing the way for the final destruction of the tree.

420. Objects of **Pruning.** If intelligently performed, pruning has one of four objects in view, viz.:

a — *To change* the *form* of the plant, as to outline or density (*formative pruning*).

b — *To stimulate development* in some special part, as to promote the growth of wood or the formation of flower-buds (*stimulative pruning*).

c — *To prevent some impending evil* to the plant, as to arrest or exclude disease (*protective pruning*).

d — *To hasten or retard maturity* (*maturative pruning*).

A — Formative Pruning

This aims to regulate the *form* of the plant with reference to *outline* or *density*, or to *strength of stem*. Pruning for *outline* includes pruning (a) for *symmetry* or *picturesqueness;* (b) for *stockiness* or *slenderness*.

421. Pruning for **Symmetry** aims to develop in the plant a head that is symmetric with reference to its trunk.

The general principle involved is the suppression of growth in all parts that tend to grow beyond the lines of symmetry (Fig. 158). This is best accomplished by pinching (416 a) during the growth period, thus economizing the plant's energy; but when the pinching has been neglected, the shoots that grow out of symmetry may be cut back during the dormant period.

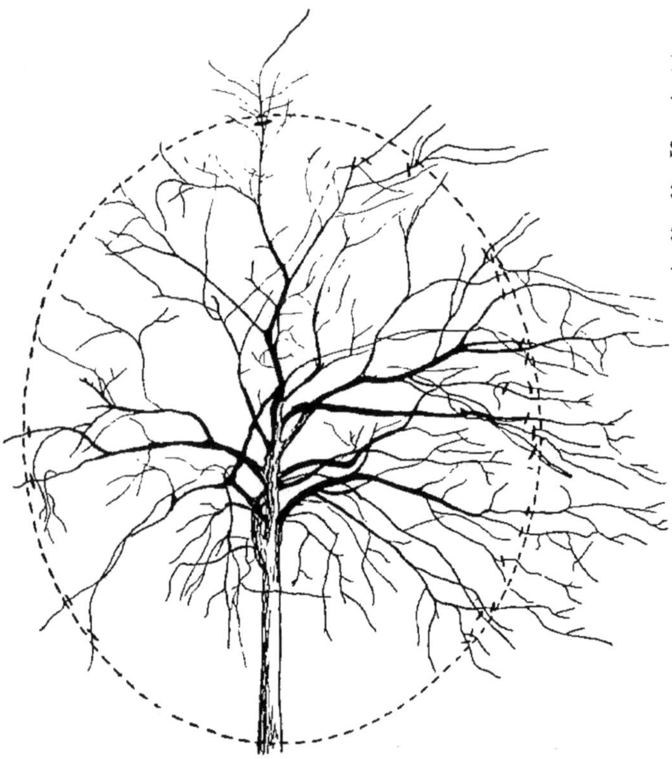

FIG. 158. Pruning for symmetry. The branebes growing beyond the ideal outline, indicated by the dotted line, should be cut off at the points indicated.

In pruning for symmetry, the plant should generally be encouraged to develop the form that is natural to the particular species or variety; e.g., the American elm tree,* which naturally develops an open, somewhat spreading head tending to be broadest toward the top, should not be pruned to the same form as the sugar maple † that develops a more roundish and compact head. Evergreens. are sometimes pruned to ideal forms, as in *topiary work,* a practice that is generally condemned by good taste.

* *Ulmus Americana.* † *Acer saccharinum.*

422. Pruning for **Picturesqueness** is seldom employed. It requires a thorough knowledge of pruning and of plant growth, combined with the conceptions of the artist.

423. Pruning for **Stockiness** aims to develop a low head with abundant branching, and a strong trunk. It is best accomplished by pinching (416 a) the uppermost growing points during the growth period, and encouraging low branching on the stem. If a spreading form is desired, the lower branches should be pruned to outside buds (Fig. 157).

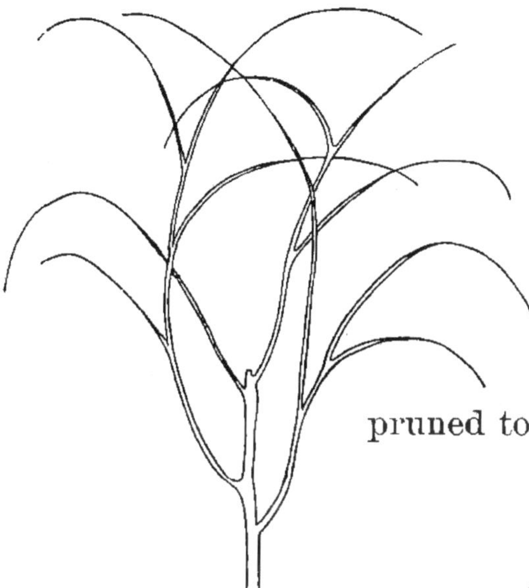

FIG. 159. Raspberry cane rendered stocky by pruning.

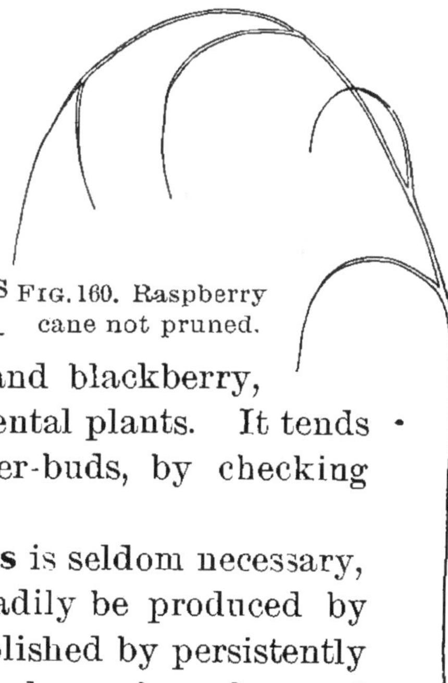

FIG. 160. Raspberry cane not pruned.

Pruning for stockiness is much practiced in the raspberry (Figs. 159 and 160) and blackberry, in hedges and in many ornamental plants. It tends to the production of flower-buds, by checking growth of wood (137).

424. Pruning for **Slenderness** is seldom necessary, as a slender growth may readily be produced by close planting. It is accomplished by persistently removing or cutting back the lower branches, and permitting only a few branches to develop near the terminus of the stem.

425. Pruning for **Density** applies either to increasing or decreasing the density of the head. In ornamental and shade trees, a compact head is often desirable, while in fruit trees, a head that admits abundant light and air (Fig. 164) is important (243). To increase density, encourage lateral branching by pinching all the more prominent terminal growing points (Fig. 162). In some coniferous trees, as the Norway spruce,* disbudding of the terminal shoots (Fig. 161) in spring is advisable, and in woody plants too tall for pinching, the more prominent terminal growing-points may be cut back with the pole shears (431), which causes the head to grow more dense.

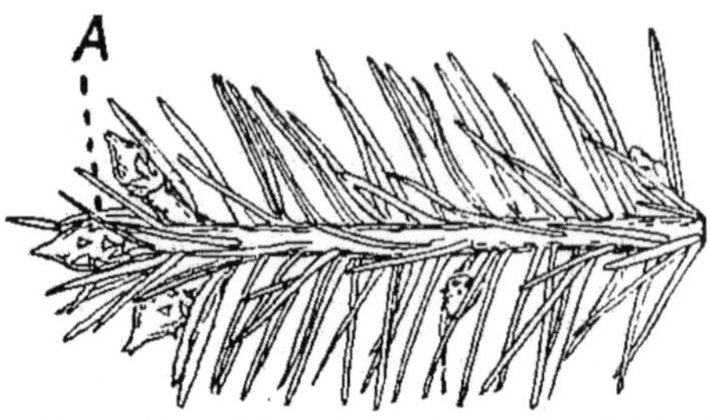

FIG. 161. Showing how to *disbud* shoots of some coniferous trees. Picking out the terminal bud A in spring usually causes both the adjacent lateral buds to develop.

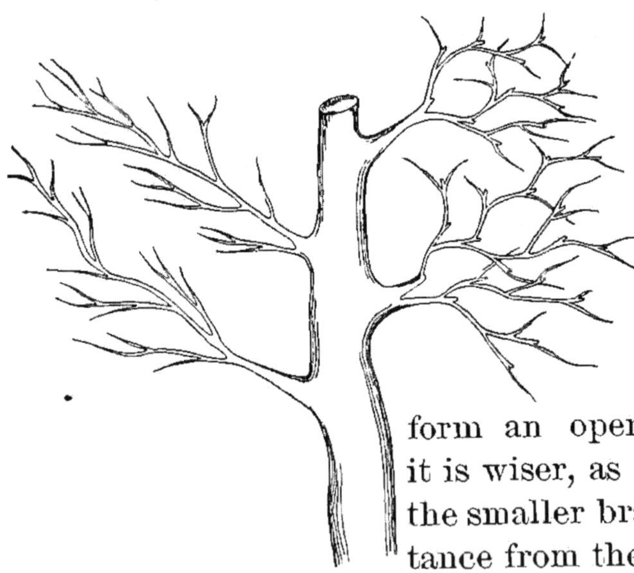

FIG. 162. Showing how density of growth is promoted (right-hand side) by persistent pinching of the terminal growing points.

In pruning to form an open head (Fig. 164), it is wiser, as a rule, to thin out the smaller branches at some distance from the trunk than to remove large branches at their union with the trunk.

426. Pruning for **Strength.** a—*of the Trunk.* Trees and plants grown in closely-

* *Picea excelsa.*

planted nursery rows often have trunks insufficiently developed to support the head, when planted by themselves. To remedy this defect, we promote the formation of new vascular bundles (68, 124) by inducing branching, which we accomplish by cutting back the top in proportion to the slenderness of the trunk (423).

FIG. 163. Unpruned apple tree, with head too dense to admit light.

b — *of the Branches.* Trees expected to support heavy crops of fruit, or to endure high winds, should have branches developed with special reference to strength. In such cases, several medium to small branches are better able to endure the strain than a few large ones (245 b), and the loss to the tree of a small branch, should it occur, is less serious than that of a large one. Forming the head of fruit trees of three or four main branches is to be discouraged for this reason. Several small branches from a common trunk are better, and these should be en-

FIG. 164. Apple tree pruned with open head, to admit abundant light.

couraged to leave the trunk at nearly right angles (Fig. 157). Forks in the trunk of fruit trees, dividing the wood into two nearly equal parts are objectionable, as one or the other part is very liable to split down under the weight of a heavy fruit crop.

Main branches inclined to split down may sometimes be prevented from doing so, by twisting two smaller branches together, to form a connection between them (Fig. 165). The branches thus twisted often grow together, forming a tie of great strength. A main branch that has actually commenced to split down may often be saved by passing an iron bolt through it and the remainder of the trunk. A bolt thus inserted may become entirely inclosed by later growth.

B — STIMULATIVE PRUNING

This depends upon the principle that *the suppression of growth in one direction tends to stimulate it in others.* Stimulative pruning may be employed either to stimulate growth of leaves, branches and roots, or of flower-buds.

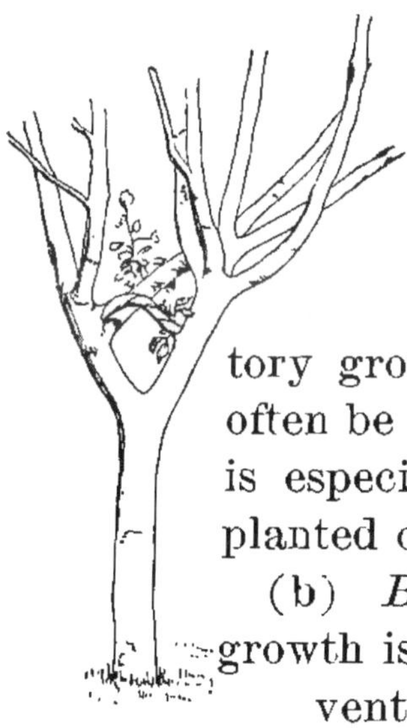

FIG. 165. Branches of fruit tree tied together by a graft formed of twisted twigs.

427. a — **Pruning** for **Growth** may be performed, (a) *By removing* a *part* of the *branches,* thus reducing the number of growing points and the surface exposed to evaporation. Plants that are not making satisfactory growth through feeble root action, may often be invigorated by this treatment, which is especially useful in trees recently transplanted or weakened by overbearing.

(b) *By suppressing reproduction.* When growth is desired, it is often advisable to prevent the development of flowers. Newly planted strawberry, raspberry and blackberry plants usually make better growth the first season if the flower-buds are picked off. The removal of flowers in the potato plant tends to stimulate the growth of tubers, especially in varieties that form seed. The removal of flower-buds from cuttings in the propagating bed encourages the formation of roots. *Topping* tobacco and rhubarb plants (416 c) causes the leaves to grow larger, and of onion plants stimulates growth of the bulbs. *De-tasseling* corn encourages growth of the ears (416 d). *Thinning* fruit on plants that incline to overbear, causes the remaining fruits to grow larger (416 i, 160).

428. b — **Pruning** for **Flowers** or **Fruit.** Since checking growth tends to stimulate the formation of flower-buds (135 b), we encourage flowering in plants that incline to luxuriant growth, by pruning which tends to check vigor. This may be accomplished,

(a) *By pinching the terminal buds* during the growth period, as is often practiced upon tardy-bearing fruit trees or upon seedling fruit trees of which it is desirable to soon learn the quality of the fruit. To be successful, it must be performed rather early in the growing season, and before the time for the formation of flower-buds. The blossoms do not usually appear until the season following the pinching.

With plants that flower at the terminal growing points of the principal branches, as the spiræas, hydrangeas, rhododendrons etc., pinching to promote flowering is not advisable; as it tends to reduce the size of the flower clusters.

(b) *By cutting back the new growth.* Woody plants that flower on stems more than one year old, as the apple, pear, currant etc., when grown on rich or well cultivated ground, or that have been too severely pruned, often tend to produce an excess of new wood with a very feeble development of flower-buds. In such cases, it is advisable to equalize the growth by a moderate cutting back of all the young shoots. This must, however, be done with judgment. If the cutting back is too severe, it will stimulate more wood growth rather than the development of flower-buds.

(c) *By root pruning.* This checks growth by reducing the number of root-tips, and thus cuts off a part of the water supply. It is applicable to the same cases as pinching, and is accomplished by cutting off the extremities of the roots by inserting the spade in a circle about the plant, or in the case of trees of considerable size, by digging a trench sufficiently deep to sever the lateral roots. The severity of the root pruning advisable will depend upon the vigor of the growth it is desired to check.

(d) *By obstructing the growth current.* This is accomplished by *ringing* (416 g), by *notching* (416 h) and by *peeling the stem* (73).

When *ringing* is practiced, the width of the belt of bark removed should usually not be so great that the wound cannot heal over the same season by the callus formed on the upper edge of the ring (80), and it must be made sufficiently early to give time for healing. A wider ring will sometimes heal if the ringing tools are not inserted deeper than the cambium layer (81). In the grape vine, in which ringing is often practiced to increase the size and earliness of the fruit, the width of the belt removed is less important, since the canes that have borne fruit are generally removed in the annual pruning. But in fruit trees, the belt of bark removed should not much exceed one-eighth inch in width. Simply cutting through the bark with the pruning saw often accomplishes the desired end.

Notching above or below a bud or twig affects it much as girdling affects the entire girdled member. Notching below a bud or twig, therefore, checks its growth, and is often followed by fruiting in that part.

Peeling the stem has sometimes been practiced to make barren trees fruitful (73). It is a hazardous operation at best, and should only be used as a last resort. It is accomplished by making two cuts around the trunk, usually several inches apart, and just through the bark, with one or more vertical cuts between them, after which the bark between the circular cuts is carefully peeled off. It should only be performed during a period of very rapid growth, and at a time when the wood is well supplied with reserve food, i. e., some time after the tree has put out leaves. It is most likely to succeed in warm, dry weather, and when the wound is not shaded after peeling; otherwise, injurious fungi are apt to infect the ruptured cells.

C — PROTECTIVE PRUNING

429. Dead or **Dying Members** of a plant **Should Be Promptly Removed,** since they more or less endanger its well-being. Dead branches of any considerable size invite decay into the stem which often results disastrously (419). Branches that are dying from infection by a fungous parasite, as the apple or pear blight, or the black knot of the plum (323), are especially dangerous and should always be removed as soon as discovered. Branches that tend to interfere with the growth of others already formed should be checked by pinching

(416 a), and those that interfere by too close contact should be cut back in proportion to the interference.

Scraping off the dead bark scales from old fruit trees tends to remove certain destructive insects or their eggs. It should be done during the growing season, and a short-handled hoe or a box-scraper is convenient for the work. Trees subject to sun-scald should generally not be scraped unless other trunk protection is given.

D — MATURATIVE PRUNING

430. a — *Pruning to hasten maturity.* This is seldom practiced. In nursery trees that tend to grow too late, and are thus subject to winter killing, the leaves are sometimes removed two or three weeks before the time when hard frosts are expected, to encourage ripening of the wood.

The later tobacco plants in a plantation are usually topped at the time the main crop is pushing the flower stalk, which causes their leaves to mature in season to be harvested with the rest of the crop.

b — *Pruning to retard maturity,* see (159).

431. The Principal Pruning Implements are the following:

The pruning knife (Fig. 166) is useful for removing small woody shoots. The blade should be of good steel, and the point should curve forward a little, to prevent the edge from slipping off the branch. The handle should be large to avoid blistering the hand, the base of the blade should be thick to furnish a support for the thumb, and the rivet should be strong enough to sustain hard pressure upon the handle.

In using the pruning knife, the shoot to be cut off should generally be pressed with one hand toward the member that supports it and the blade should be inserted at the proximal side. Care is necessary to prevent the blade from cutting too far.

The pruning saw (Fig. 167) is useful for cutting off large limbs. Two toothed edges are preferable to one, as the second edge tends to prevent "pinching." It is well to have the teeth on one edge point backward, as this enables the saw to cut either when pushed or pulled. Sometimes the blade is curved like a sabre, with the teeth on the concave edge pointing backward. The blade should taper nearly to a point, to enable it to enter between crowded branches.

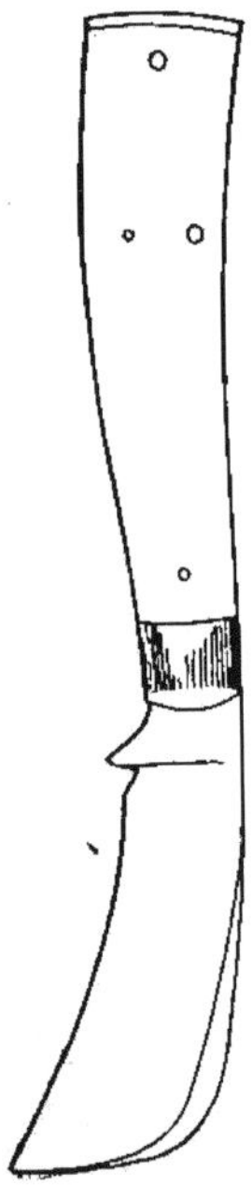
FIG. 166.

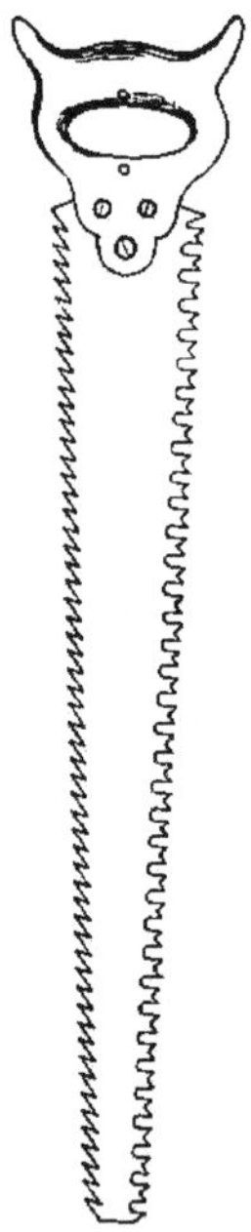
FIG. 167.

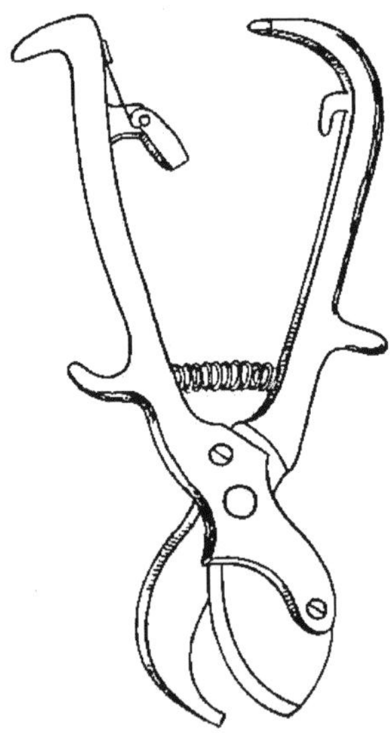
FIG. 168.

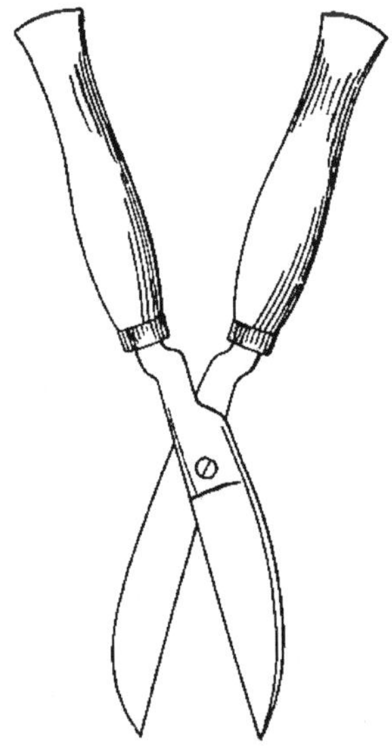
FIG. 169.

FIG. 166. Pruning knife. FIG. 167. Pruning saw.
FIG. 168. Pruning shears. FIG. 169. Hedge shears (much reduced).

The pruning shears (Fig. 168) may be used for the same purpose as the pruning knife, but they cut less smoothly, and less close to the supporting member. They should be used with the beveled edge of the blade in close contact with the supporting member. They are excellent for cutting cions (386), and making cuttings (358). The form shown in the figure is perhaps the best one extant.

The hedge shears (Fig. 169) are especially useful for pruning hedges.

The lever shears (Fig. 170) are useful for cutting off sprouts about the base of trees.

The pole shears (Fig. 171) are useful for cutting back the shoots of tall trees, and for removing sap sprouts (224), though for this purpose they have the fault of the pruning shears in not cutting sufficiently close to the branch. They should not be used for shoots much exceeding one-half inch in diameter.

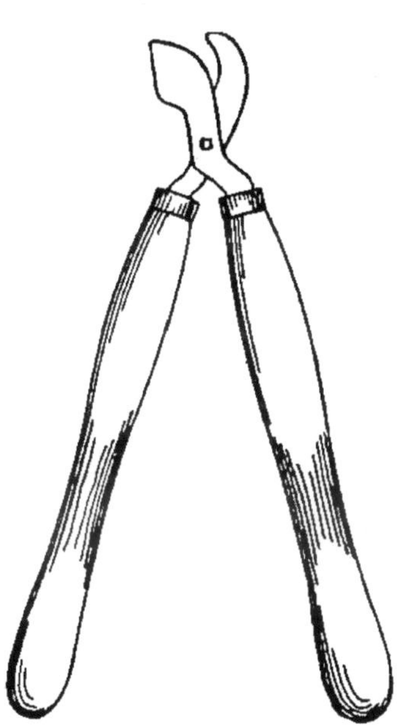

FIG. 170.

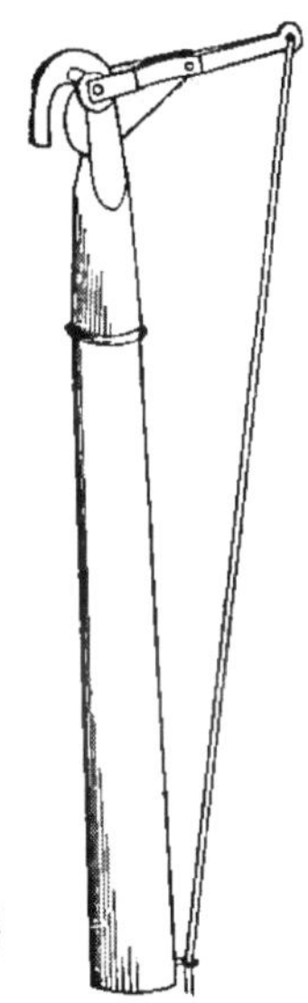

FIG. 171.

FIG. 172.

FIG. 170. Lever shears (much reduced).

FIG. 171. Pole shears. The wire connects with a lever not shown in the figure.

FIG. 172. Raspberry hook.

The raspberry hook (Fig. 172) is used for cutting off the dead fruiting canes of the raspberry and blackberry. The cutting part is made of a rod of good steel, five-sixteenths inch in diameter, flattened and curved as shown, with a moderately thin edge on the concave side of the curve. The handle should be about three feet long.

The following books are recommended for reading in connection with the preceding chapter: The Nursery Book, Bailey; Greenhouse Construction, Taft; Barry's Fruit Garden, Barry; The Art of Grafting, Baltet; The Pruning Book, Bailey; How to Make the Garden Pay, Greiner.

CHAPTER V

PLANT BREEDING

432. Plants Have Improved Under Culture. From our point of view, our cultivated varieties of plants are superior to their wild prototypes. The strawberries of our gardens are larger, more productive and firmer than those of the fields; the cultivated lettuces are more vigorous, more tender and milder in flavor than wild lettuces; and the cultivated cabbages and cauliflower are greatly superior, in the food products they furnish, to their progenitor. The superior qualities of long-cultivated plants, as compared with their wild parents, is conspicuous whenever the wild form is known.

433. Whence this **Improvement?** It probably results from two causes. a — In culture, the natural hindrances to development are largely removed. Cultivated plants are less crowded by too-near neighbors than wild plants, and they commonly receive more abundant food and moisture. They are, therefore, able to reach higher stages of development than is possible in nature, where plants are constantly restricted by environment.

b — The principle of selection has doubtless been more or less operative since the beginnings of culture (19). All of our cultivated plants must have existed originally in the wild state. The most satisfactory plants of any desirable species have been most carefully guarded, and when the art of propagation became known, these plants were most multiplied. In each successive generation, the most desirable individual plants of each species were protected and multiplied, or at least were permitted to perpetuate themselves. Since the offspring tends to resemble the parent (18), the persistent propagation from the best has resulted in more or less marked improvement. Chance crossings have aided the process (445). These facts furnish hints for the further improvement of plants.

434. The Variability of plants **Renders** their **Improvement Possible.** In a species of which the individual plants are all practically alike, as in many wild plants, we can do little in the way of plant breeding, except to give treatment that promotes variability. In a species in which the individuals manifest different qualities, however, we may hope to secure improvement by using the most desirable plants as parents from which to secure still further variability.

435. Variations are **Not Always Permanent.** If we find a chance seedling of the wild blackberry, for example, that has remarkably fine fruit, the plants grown from seeds of this fruit are not always equal in quality to the parent. The tendency, in such cases, is for the seedling plants to *revert* or go back to the ordinary type of the species, and the more marked the variation, the stronger is the tendency to reversion.

436. How to **Fix Desirable Variations.** A *fixed* variation, i. e., a variation of which the progeny resembles the parent in all important characters, becomes a *variety* (21)* as this word is used with reference to cultivated plants.

* Varieties that reproduce their more important characters when grown from seed, are often called *races*.

There are two possible ways of fixing a desirable variation:

a — *By propagating the plant by division* (345). This enables us to maintain a given variation through many generations with comparatively little deviation from the form with which we started (341). Our varieties of fruits, potatoes, geraniums and many flowering plants, and of many of our finest ornamental trees and shrubs, are fixed in this manner. It is well known that varieties propagated in this way rarely "come true" from *seed,* i. e., their seed does not usually produce plants of the same variety as the parent. But it is not practicable to propagate all plants by division.

With plants more conveniently propagated from seed, as the cereals, Indian corn and most garden vegetables, we may fix varieties to a certain extent,

b — *By persistent selection toward an ideal type.* For example, if we discover a single pea plant in a row of peas that produces earlier pods than any other plant, and we desire to fix this variation, we would save all the peas from this plant and sow them the next spring. Most of the plants from this seed will probably be later than the parent, but two or three of them may equal it in earliness. We would save the seeds from the earliest plant again, and continue this selection through several seasons. It would be well to note the incidental characters of the earliest plants, i. e., whether the pods are borne singly or in pairs, if they are straight or crooked, and whether the plants are tall or dwarf. Having decided on the characters that seem to accompany the extreme earliness, we should save seeds only from plants that show all these characters. After following this kind of selection eight or ten years, we may be able to introduce a new variety of pea.

It is impossible to so fix variations in plants grown from seed that they will continue to come true without a certain amount of selection, hence varieties propagated by seed continually tend to "run out," i. e., to lose their distinctive characters. Seed growers find it necessary to use the utmost care in maintaining their varieties, and the more distinct a variety propagated by seed, the more difficult it is to maintain.

437. Seed Selection is of **Great Importance.** From what has been said, it is clear that the cultivator cannot afford to be indifferent as to the quality of the seed he sows. It is not enough that the seed is fresh and plump; it should be of carefully-bred varieties. In the cabbage and cauliflower, success or failure in the crop will depend largely upon the quality of seed sown, and the same is more or less true in all crops grown from seed.

438. We Can Induce Variation, in some cases, by special treatment of the parent plants, or by the use of a particular selection of seed.

a — *By culture.* It is generally conceded that culture tends to promote variations that would not have appeared in the wild state, in consequence of the changed growth conditions. In improving wild plants, therefore, we probably have a better chance of securing variation by gathering seeds from such wild plants that have been placed under high cultivation than from those that have not been submitted to culture.

b — *By growing seedlings.* In plants habitually propagated by division (345), as the apple, potato, dahlia etc., we secure variation by growing plants from seed. The parent plant, not having been fixed by long selection, as is the case with varieties grown from seed, is in a *state of variation*, and hence its progeny usually varies widely. From these variable seedlings, desirable individuals may be selected for fixing. Since most of our varieties that are propagated by division are highly developed, their seedlings are usually, though not necessarily, inferior to the parents.

c — *By crossing varieties or species.* This is the most important method of plant improvement. By procuring fecundation of the germ cell of a plant of one variety with pollen from a plant of a different variety or species (150) through cross-pollination (152), we obtain a variable progeny of which the individual plants may be expected to resemble both parents in different degrees. For example, if we secure fecundation of a number of ovules of the Worden grape with pollen from the Delaware grape, and plant the seeds from the fruits thus secured, we may expect that some of the seedlings will resemble both parents about equally, that others will chiefly resemble the Worden, but will show a few characteristics of the Delaware, while others again will chiefly resemble the Delaware, but will possess a few characteristics of the Worden. It would not be surprising if we secure a vine having the vigor, productiveness and large fruit of the Worden, with the color and delicious flavor of the Delaware. This we may almost certainly accomplish if we continue our trials a sufficient time. In other words, *we may often combine the good qualities of two varieties into a single variety by securing a number of cross-fecundations be-*

tween the two (440), *and rearing plants from the seeds thus formed.*

439. The Selection of Subjects for Crossing. If the object of crossing is simply to secure variation, as is sometimes the case with wild fruits, the parents should differ from each other as widely as possible, provided only that they are capable of crossing freely. Crosses between allied species (hybrids (23)), when this is possible, will be more likely to accomplish the object sought than between plants of the same species.

If the object is the improvement of present varieties, the parents should be chosen with reference to the qualities desired in the new variety. For example, if it is desired to produce a hardy, late-keeping apple, of first quality, any hardy variety that keeps well, whatever its quality, may be crossed with any other hardy apple of first quality, whether it keeps poorly or well, though of two apples of first quality, the better keeper should be chosen.

The plant breeder should first have a definite idea of the qualities he desires to secure in his proposed variety, and should then study with much care the qualities of the varieties that he proposes to use as parents. The two varieties that contain the largest number of the desired qualities should be chosen.

440. Cross-Fecundation is accomplished through cross-pollination of the flowers (152); i. e., by placing pollen from the anthers of a flower of one of the varieties we desire to cross upon the stigma of the other variety.

441. Preparing the Flower for Crossing. To prevent self-pollination (152) in perfect flowering plants (154), we *emasculate* (e-mas′-cu-late) the flowers, i. e., remove the anthers (144) before the pollen is mature. Prior to maturity, the anthers are generally pale in color and nearly smooth on the surface, with no visible pollen, but a little later, the pollen in most plants is visible as a bright yellow dust adhering to the anthers. The anthers may be picked off with the forceps, or the filaments that support them may be clipped off with the points of the scissors.

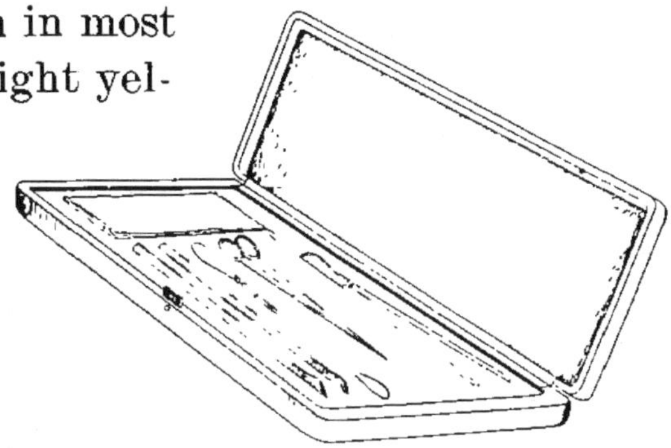

FIG. 173. Case of instruments and sacks for crossing plants.

They must generally be removed before the petals open (143). The latter may be gently opened with the forceps or needle, or they may be carefully removed.

In the flowers of certain plants, as the pea, wheat and grape, pollination takes place before the blossom opens, hence in these plants it is necessary to emasculate the flowers very early.

FIG. 174. Emasculated flower inclosed in sack.

442. To Prevent Undesired Pollination, the blossom should be inclosed by tying over it a sack of thin cloth or paper at the time of removing the anthers. The sack will of course have to be removed for pollination, after which it should be promptly replaced.

Pollination should be performed twenty-four to forty-eight hours after emasculation (441), the period depending upon the plant and the stage of development of the flower at the time of the latter operation (151). Applying the pollen on two consecutive days tends to insure success.

The pollen is applied by placing an anther (144) containing mature pollen in direct contact with the stigma (145), or by removing some of the pollen upon the back of the point of a penknife or by means of a camel's-hair brush, and carefully applying it to the stigma. A pin, of which the head has been flattened by hammering, inserted in the end of a stick, forms a convenient tool for this work.

The best time for pollination in the open air, is often in the early morning, since the atmosphere is then usually still, and contains little pollen from other flowers, which, if freely present in the air, may vitiate the results of the pollination.

443. The After-Care of Crosses. After the last pollination, the blossom should again be inclosed until fecundation is effected, which is indicated by a rapid enlargement of the ovary. The paper sack may then be replaced by one of mosquito netting. This should be securely, but not too tightly, tied about the stem of the pollinated flower, to protect the inclosed fruit or seed-vessel from injury during growth and maturity, as well as to render it conspicuous. A label should be placed within the sack, or tied on with it, giving the name of the variety whence the pollen was secured. It is also desirable to record all the operations and observations relative to the crossing.

444. The Selection of Crossed Seedlings is a most important operation in producing new varieties by crossing. If none of the seedlings of the first generation exhibit the desired qualities, those of a succeeding generation may exhibit them. The plants nearest the ideal should be selected, and all the seeds from these preserved for planting. When the ideal plant is found, it may be readily fixed by means of cuttings or grafts in plants generally propagated in this way. In those propagated by seed, several generations of culture and selection may be necessary before the progeny will uniformly resemble the parent.

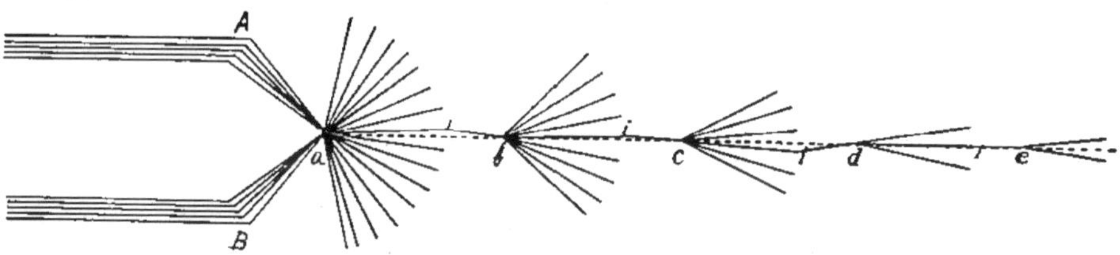

FIG. 175. Diagram illustrating the selection of seedlings from a cross.

The variations in the seedlings from two crossed varieties, and the kind of selection needed to fix the desired variation, are illustrated by the following diagram (Fig. 175). Let *a* represent the seeds from two crossed flowers *A* and *B*. The plants from these seeds will probably be quite variable, as is indicated by the divergent lines. Let us suppose the variation marked *i* to be nearest the ideal form. The plants grown from *i* will again be quite variable in the second generation *b*, but probably less so than in the first generation. No plants of the second generation may be nearer the ideal type than those of the first generation, but we select the plant nearest to our ideal, and plant the seeds from this. Each succeeding generation may be expected to produce less of variability than the one before it. By and by, we may hope to secure a form that approaches our ideal and comes tolerably true from seed.

445. Planting with **Reference** to **Chance Crossings.** Many valuable varieties have unquestionably arisen from accidental crosses between plants of different varieties that chanced to be growing in proximity. Profiting by this hint, varieties are sometimes planted near together to favor self-crossing, a practice to be encouraged.

446. Those Who Improve Plants are **True Benefactors.** He who produces fruits or flowers for others works a transient good. But he who produces a variety of fruit or flower that is superior to any now known confers upon his race a *permanent* good. Until the introduction of the Wilson strawberry, the markets of our country were not supplied with this delicious and wholesome fruit, because no known variety was sufficiently productive to be generally profitable, or sufficiently firm to endure long carriage. What a blessing was conferred upon us by a Mr. James Wilson, of Albany, N. Y.! There are wild fruits in our copses to-day that are doubtless worthy of improvement, and in most of our fruits now under culture the development of superior varieties would greatly enhance their value. "The harvest truly is great, but the laborers are few."

The following books are recommended for reading in connection with the preceding chapter: Plant Breeding, Bailey; Variations of Animals and Plants Under Domestication, Darwin; Propagation and Improvement of Cultivated Plants, Burbridge; Origin of Cultivated Plants, De Candolle.

APPENDIX

A SYLLABUS OF LABORATORY WORK

The laboratory exercises here outlined have been used by the author in his instructional work.

Each student performs the exercises, so far as possible, and the apparatus needed is provided. The student should be required to write a description of the work performed, stating results in every case, supplementing his notes by drawings in special cases.

It has not been found practicable to make the lecture room and laboratory work fully correspond as to time, but the effort has been made to do this as far as possible.

A greenhouse is very desirable for this kind of instruction, and if the instruction is given in winter, a "garden house," i. e., a glass house inclosing an unobstructed area of garden soil is scarcely less importaut. But in the absence of these conveniences, a few window boxes will furnish a tolerable substitute.

When the exercises are carried out during winter, considerable foresight is essential to have the needed materials in condition for use at the proper time.

To stimulate observation (1).* A few object lessons are given to encourage observation and correct reasoning. A twig, an ear of corn or a potato tuber is given to each student and all are encouraged to vie with each other in discovering new points, and in discussing the reasons therefor. This lesson is frequently repeated during the term.

* The numbers in parenthesis refer to the paragraphs in the book.

Cell structure (12). The students examine the pulp of a mealy apple and of a potato, and cross-sections of a young bean plant, with simple lenses of rather high magnifying power. If a compound microscope is available, many mounted objects illustrating the cell structure of plants may also be shown.

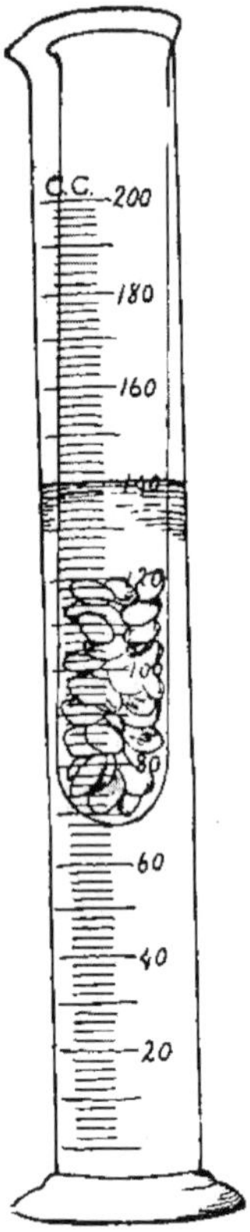

FIG. 176. Device for weighing and determining the volume of seeds.

Absorption of water by seeds (26). For the exercises suggested by paragraphs 26 and 27, a means of weighing and of measuring the volume of large seeds, as beans, with some degree of accuracy is needed. The device shown in Fig. 176 answers this purpose, and one can be provided for each pair of students at a moderate cost. It consists of a graduated glass cylinder of 200 cubic centimeters capacity and a test tube about 6 inches long. For determining the volume, the cylinder is partly filled with water and the height to which the water rises is noted. The seeds are then dropped in and the glass is shaken a little to remove the air bubbles. The height of the water is again noted, when the difference in the two readings indicates the volume of the seeds in *cubic centimeters.* For weighing, the empty test tube is placed in the cylinder in the position shown (Fig. 176). The height to which the water rises is then noted, after which the seeds are dropped into the test tube, and the top of the cylinder is jarred slightly by tapping it with the pencil. The height of the water is again noted, when the difference in the readings indicates the weight of the seeds in *grammes.*

The test tube should float in the center of the cylinder, as shown, and the readings should be taken with the eye on a level with the surface of the water.

Each student (or pair of students) is provided with the apparatus shown in Fig. 176, and with two bottles of at least 100 cc. capacity, with corks. Each bottle should have a strip of white paper pasted vertically upon it to receive the name of the student and other data.

Each student weighs or measures the volume of 50 fresh seeds of the bean, pea or Indian corn in the manner described above. Having noted the weight or volume in his notebook, he pours the seeds, with the water, into one of his bottles, corks the latter and writes his name, with the date, on the paper pasted on its side. He then repeats the process with seeds of the honey locust, yellow wood or some other seed that does not readily absorb cool water, and after recording the data in his notebook, places the bottles in a warm place until the following day, when he again determines the weight or volume of the two kinds of seed. The seeds placed in the first bottle will usually be found to have nearly or quite doubled in size, while those in the second bottle have scarcely swollen at all.

Next, show the class a sample of the second lot of seeds that have fully swollen from soaking in hot water. Impress upon their minds the fact that while most seeds readily absorb moisture at ordinary temperatures, a few kinds do not, and seeds of the latter class need to be soaked cautiously, before planting, in hot water (27*d*).

The rate at which seeds absorb water depends

a — *Upon the water content of the medium* (27). Weigh 3 samples of navy beans. Place one sample in water, a second in very damp earth and the third in slightly damp earth. Weigh again the next day and compute the water absorbed by the three lots.

b — *Upon the points of contact.* Weigh 2 samples of navy beans, placing one sample in moist soil without compacting, and the second in the same kind of soil well compacted about the seeds. Determine the water absorbed by the two samples the next day.

c — *Upon temperature.* Repeat the above with 2 samples of navy beans, placing one lot in a temperature of 80° to 90° F., and the other in 40° to 50° F.

Other means of using the apparatus shown in Fig. 176 will occur to the thoughtful teacher. It may be used for determining specific gravities by dividing the weight by the volume.

Germination (28). Give an exercise in testing seeds with the apparatus shown in Fig. 6.

Moisture essential to germination (29). Soak one lot of navy beans in water until they are fully swollen, and another lot until they are about half swollen. Wipe the beans as dry as possible, put each lot into a bottle, cork the bottles, and set them in a warm room. The fully-swollen beans will usually germinate promptly, while the others will not.

Oxygen essential to germination (31). Perform the saucer experiment as described.

Also place seeds of rice in two bottles, and add to each, water that has been boiled 20 minutes; cover the water in one bottle with a little olive- or cotton-seed oil. It is important to soak the seeds a short time in boiled water before putting them in the bottles to remove the air in contact with their seed-cases.

Germination hastened by soaking seeds (36). Soak seeds of Indian corn two or three hours in warm water, and let each student place in a seed tester a sample of the soaked seeds, with one or two other seeds of the same kind that have not been soaked.

Germination hastened by mutilating the seed-case (37). This may be illustrated with seeds of the navy bean, in the seed tester.

The plantlet (41). Place seeds of radish, onions etc., loosely on the surface of a saucer filled with fine moist loam; keep the surface moist and note the repeated attempts of the hypocotyl to enter the soil.

Seeds of the pumpkin family should be planted flatwise (43). Plant seeds of the pumpkin or squash, in the three positions indicated, in large greenhouse saucers. Cover each saucer with a pane of glass and place all in a warm room until the plantlets appear, after which note the number of each lot of seeds of which cotyledons appear above the surface.

Development of plantlets (45-47). Devote several exercises to a study of the development of plantlets of the bean, pea, wheat, Indian corn, pumpkin, etc. To furnish the plantlets, seeds of the different sorts should be planted on several successive days, beginning at least 10 days in advance.

Not all seeds should be deeply planted (48). Plant seeds of the bean, pea, Indian corn and wheat in 6-inch flower pots, at three different depths, viz., ½ inch, 3 inches and 6 inches from the bottom; place the pots in a warm place for 3 weeks, after which carefully remove the soil, noting the germination of the seeds in the different layers.

Vigor of plantlet proportionate to size of seed (49). Plant large and small specimens of navy beans by themselves, in greenhouse saucers, and permit them to germinate. The smaller seeds usually germinate earlier than the larger, but they produce more slender plantlets, which soon fall behind the others in development.

Plantlet visible in the seed (54). Boil samples of various kinds of seeds until they are fully swollen, after which require the students to dissect them and to seek out the plantlets. Lenses, needles and forceps are very useful in this work.

The cotyledons a storehouse for food (60). Remove the cotyledons of some bean plantlets growing in a flower pot or saucer, leaving those of other plantlets intact. After a week note the result in the checked growth of the mutilated plants.

Vascular bundles (68). Study these as shown in the stalk of Indian corn, in the leaf stems of various plants and in the leaf scars on the stems of plants.

Cambium layer (69). Locate this in sections of various dicotyledonous stems, including the potato tuber; also note the absence of the cambium layer in monocotyledonous stems.

Root-hairs (101). Study these as illustrated when seeds germinate in the seed tester. Germinated radish-seeds, left in the seed tester two or three days, usually develop root-hairs in great abundance. Also search out the root-hairs in potted plants. Emphasize the difference between root-hairs and root branches.

Effects of transplanting on root branching (105). Study young plants of lettuce, tomato, cabbage etc., that have been pricked off, and compare their roots with those of others that have not been pricked off.

Relation of roots to food supply (112). Plant seeds of the radish in saucers containing clean sand and potting soil respectively, and when the seedlings have attained some size, wash out and examine the roots in the two soils.

Root tubercles (113). Study the roots of young clover plants of various ages, and note how early in the development of the plant the tubercles are discernible.

Underground stems (115). Study the development of the potato plant from growing specimens, noting the points at which the tuber-bearing stems originate, and the marked difference between these and the roots.

Nodes and internodes (116). Observe the nodes in the stems of many plants, noting the relation of the diameter of the young stem to the length of the internodes; also note the undeveloped internodes near the terminus of the stem.

Buds (128). Study specimens of leaf-buds from many plants, noting their structure, position etc.

Flower-buds (133). Study the form and location of the flower-buds in many plants, particularly in fruit trees.

Parts of the flower (141). Study the parts of the flower, explaining the function of each part.

Perfect and imperfect flowers (154). Study these as produced by several different plants, particularly of the strawberry.

Degree of maturity necessary to germination (163). Test seeds of Indian corn, pea, tomato etc., that were gathered at varying stages of maturity.

Seed vitality limited by age (165). Test seeds of lettuce, parsnip, onion etc., 1 year, 2 years and 5 years old respectively.

Stratification of seeds (170). Perform the process, as described, in boxes or large flower pots.

Sun-scald (186). Require each student to make a lath tree protector (Fig. 59).

Winter protection of plants (202). Protect half-hardy shrubs by wrapping them with straw or covering them with earth.

Foretelling frost (207). Devote an exercise to the use of the psychrometer and the computation of the dew point.

Plant protectors (279). Require each student to make at least one plant protector, as shown in Fig. 67, patterns for which are to be furnished.

Kerosene emulsion (294). Let each student make a given quantity of the kerosene emulsion after one of the formulas given.

Spraying pumps (304). Give at least one exercise to the construction and use of spraying pumps and nozzles.

Prevention of grain smuts (325). Require the students to treat a quantity of oats to hot water as described.

Bordeaux mixture (329). Require each student to make a stated quantity of the Bordeaux mixture after the formula given.

Propagation by seeds (344). Instruct the students in the use of the hand seed-drill and broadcast sower. Let them ascertain how much clover seed the broadcast machine is sowing per acre, by laying on the ground or floor several sheets of paper, exactly one foot square, painted with glycerine to catch the falling seeds. Having learned the average number of seeds deposited per square foot with a given rate of motion of the machine, let the students compute the number of seeds sown per acre, and reduce this to ounces. The number of clover seeds in an ounce may be ascertained by dividing an ounce of the seed among the students for counting.

Propagation by layers (349). Instruct the students in layering canes of the grape, and in mound-layering the stems of the gooseberry.

The bulb (352). Dissect bulbs of the onion, tulip, lily, etc., ascertaining their structure and finding the embryo flowers.

The cold-frame (364). Require the students to make a drawing and write a description of a cold-frame, from a model furnished them.

The hotbed (365). Let the students assist in making a hotbed after the plan described. Also let them note the temperature of the soil within the frame on several successive days after the bed is finished, and give them instruction in ventilating the hotbed.

The propagating bed (368). Require the students to make a propagating bed in the greenhouse, after the plan described.

Stem cuttings (373-375). Let the students make cuttings from the stems of the grape, currant, etc., and plant them, both in the propagating bed and in the garden.

Root-cuttings (376). Give a lesson in making root-cuttings of the raspberry or blackberry, in packing the same for winter storage, and in planting them in the propagating bed, and in the garden.

Green cuttings (380–381). Give a lesson in making and planting cuttings of coleus, geranium, rose etc., followed by instructions in the care of green cuttings in the propagating bed.

Leaf cuttings (382). Give a lesson in making and planting leaf cuttings of the begonia.

Grafting wax etc., (387–389). Give a lesson in making grafting wax, grafting cord and grafting paper, as described.

Whip-grafting (390–391). Give several lessons in whip-grafting including grafting both of the stem and of the root.

Cleft grafting (392). Give one or two lessons in cleft grafting.

Side grafting (393). Give a lesson in side grafting, as described.

Budding (394). Give one or more lessons in budding, as described. The bark on the stocks may be made to peel by boiling, and trimmed bud sticks may be preserved for winter use, in dilute alcohol.

Approach grafting (399). Give one exercise in approach grafting as described.

Packing plants for transportation (405). Devote one exercise to packing strawberry, cabbage or some other herbaceous plants, as described.

Heeling-in, Replanting (408–410). Give one or more lessons in heeling-in and planting trees, as described; also at least one lesson in planting root grafts, cuttings and herbaceous plants as shown in Figs. 143–144; and a lesson in planting strawberry plants, as shown in Figs. 145–146.

Potting and shifting (412). Give two or more lessons in potting and shifting, as shown in Figs. 147–150.

Pruning (427 etc.). Give one or more lessons in pruning by the methods described.

Cross pollination (441). Give one or more lessons in cross pollination, as described.

INDEX.

The numbers refer to pages.